# 30th International Conference on Adaptive Structures and Technologies (ICAST 2019)



Montreal, Canada
7 - 10 October 2019

**Editors:**

**Ramin Sedaghati**
**Yong Chen**

ISBN: 978-1-7138-0794-0

**Printed from e-media with permission by:**

Curran Associates, Inc.
57 Morehouse Lane
Red Hook, NY 12571

**Some format issues inherent in the e-media version may also appear in this print version.**

Editors: Ramin Sedaghati
Concordia University
Montreal, QC, Canada

Yong Chen
National Research Council (NRC)
Ottawa, ON, Canada



**Additional copies of this publication are available from:**

Curran Associates, Inc.
57 Morehouse Lane
Red Hook, NY 12571 USA
Phone: 845-758-0400
Fax: 845-758-2633
Email: curran@proceedings.com
Web: www.proceedings.com

Proceedings of

# $30^{th}$ International Conference on

# Adaptive Structures and Technologies

# (ICAST 2019)

October 7-10, 2019

Concordia University

Montreal, QC, Canada

Chairs: Ramin Sedaghati

Yong Chen

Editors:

Ramin Sedaghati
Concordia University
Montreal, QC, Canada

Yong Chen
National Research Council (NRC)
Ottawa, ON, Canada

# Preface

Welcome to Montreal, Canada, and welcome to the 30th International Conference on Adaptive Structures and Technologies (ICAST2019).

ICAST is the longest-running international conference on adaptive, intelligent, or smart materials and structures. The conference convenes every year and the venue rotates between the Americas, Asia and Europe since its inception in 1990 with the US-Japan Conference on Adaptive Structures in Hawaii. Following the successful ICAST conference in Ottawa (ICAST2007), this is the second ICAST event held in Canada.

ICAST aims at promoting research, development and application of adaptive structures and technologies through the exchange of scientific results and insight from leading international scholars and specialists as well as the young researchers. The conference showcases the state of the art in this innovative and multidisciplinary area. It provides a forum for discussion of recent advances in the fields of smart materials and structures, and encourages the transfer of advanced scientific results from research to application. The main topics include functional materials and devices, modeling and characterization, bio-inspired sensors and actuators, morphing technologies, structural health monitoring, vibration damping and energy harvesting, etc. ICAST maintains the tradition of a single-session format, and this intimate setting has furthered international understanding and launched long-lasting friendship. This year we look forward to welcoming over 90 participants who contributed 83 technical papers that are presented to you through oral presentations.

We are pleased to acknowledge the support from National Research Council Canada, Concordia University and the Canadian Society of Mechanical Engineering. The assistance from the members of the International Organizing Committee (IOC) and the Local Organizing Committee is much appreciated.

We are honored to host the 30th anniversary event of the ICAST this year, and are delighted to welcome you to the metropolitan city of Montreal, the largest bilingual city in North America, and the cultural capital of Quebec. With your participation, we anticipate a valuable, enjoyable and memorable event. We wish you a great stay in this fascinating place!

Ramin Sedaghati (Concordia University) and Yong Chen (NRC)

ICAST2019 General Chairs

# Conference Organization

## Local Organizing Committee

| | |
|---|---|
| General Chair | Ramin Sedaghati<br>Concordia University |
| | Yong Chen<br>National Research Council (NRC) |
| General Co-Chair | Subhash Rakheja<br>Concordia University |
| | Viresh Wickramasinghe<br>National Research Council (NRC) |
| Technical Chair | Masoud Hemmatian<br>Concordia University |
| Technical Assistant | Armin Rasooli<br>Concordia University |

## International Scientific Committee

| | |
|---|---|
| Haim Abramovich | Technion Israel Institute of Technology<br>Israel |
| Andrea Bergamini | EMPA/ETH Zürich<br>Switzerland |
| Diann Brei | University of Michigan<br>USA |
| Manuel Collet | CNRS<br>France |
| Antonio Concilio | CIRA<br>Italy |
| Flavio Campanile | EMPA<br>Switzerland |
| Yong Chen | National Research Council<br>Canada |
| Inderjit Chopra | University of Maryland<br>USA |
| Roeland De Breuker | Delft University of Technology<br>Netherlands |

| | |
|---|---|
| Jean-François Deü | Conservatoire National des Arts et Métiers<br>France |
| Paolo Ermanni | ETH Zurich<br>Switzerland |
| Alper Ertürk | Georgia Tech<br>USA |
| Alison B. Flatau | University of Maryland<br>USA |
| Farhan Gandhi | Rensselaer Polytechnic Institute<br>USA |
| Paolo Gaudenzi | University of Rome<br>Italy |
| Nam Seo Goo | Konkuk University<br>Republic of Korea |
| Zafer Gürdal | Skolkovo Institute of Science and Technology<br>Russia |
| Daniel Guyomar | INSA<br>France |
| Jae-Hung Han | Korea Advanced Institute of Science and Technology<br>Republic of Korea |
| Kazuyuki Hanahara | Iwate University<br>Japan |
| Holger Hanselka | Fraunhofer Institute<br>Germany |
| Daniel J. Inman | University of Michigan<br>USA |
| Xin Xiang Jiang | Canadian Space Agency<br>Canada |
| Dimitris C. Lagoudas | Texas A&M University<br>USA |
| Chih-Kung Lee | National Taiwan University<br>Taiwan |
| I. Lee | Korea Advanced Institute of Science and Technology<br>Republic of Korea |

| | |
|---|---|
| Jinsong Leng | Harbin Institute of Technology<br>China |
| George A. Lesieutre | Pennsylvania State University<br>USA |
| Wei-Hsin Liao | Chinese University of Hong Kong<br>China |
| Marcias Martinez | Clarkson University<br>USA |
| Hans Peter Monner | German Aerospace Center (DLR)<br>Germany |
| Michihiro C. Natori | JAXA<br>Japan |
| Fred Nitzsche | Carleton University<br>Canada |
| Roger Ohayon | Conservatoire National des Arts et Métiers<br>France |
| Hiroshi Okubo | Osaka Prefecture University<br>Japan |
| Jinhao Qiu | Nanjing University of Aeronautics and Astronautics<br>China |
| Dimitris A. Saravanos | University of Patras<br>Greece |
| Wieslaw Jerzy Staszewski | AGH University of Science & Technology<br>Poland |
| H. Tanaka | National Defense Academy of Japan<br>Japan |
| Marcelo Areias Trindade | University of São Paulo<br>Brazil |
| Hornsen Tzou | Zhejiang University<br>China |
| Dai Hua Wang | Chongqing University<br>China |
| Kon-Well Wang | University of Michigan<br>USA |

| | |
|---|---|
| Jean-Sébastien Plante | Université de Sherbrooke<br>Sherbrooke, Québec |
| Subhash Rakheja | Concordia University<br>Montréal, Québec |
| Ion Stiharu | Concordia University<br>Montréal, Québec |
| Ramin Sedaghati | Concordia University<br>Montréal, Québec |
| Afzal Suleman | University of Victoria<br>Victoria, British Columbia |
| Daniel Therriault | Polytechnique Montréal<br>Montréal, Québec |
| Viresh Wickramasinghe | National Research Council (NRC)<br>Ottawa, Ontario |

# Table of Contents

## Smart Structural Systems



## Shape Memory Materials

## Integration of Sensors and Actuators

## Morphing Structures

## Bioinspiration and Bioengineering



## Structural Health Monitoring

## Electro/Magneto Sensitive Materials

## Multifunctional Materials and Composites

## Energy Harvesting

# Smart and Adaptive 4D Materials – Hierarchical Manufacturing

**Hani Naguib**

*Department of Mechanical and Industrial Engineering, University of Toronto, Toronto, ON M5S 3G8, Canada*

*naguib@mie.utoronto.ca*

**Abstract**

Smart materials related research and applications have grown tremendously in the past couple of decades. The emergence of natural and synthetic smart materials possessing a set of desired properties, along with the development of novel processing techniques and a willingness of competitive manufacturers to adopt new materials has resulted in new smart materials components replacing the roles of traditional materials. Smart materials can also offer several benefits as components in adaptive and multifunctional materials. This seminar presents the design and hierarchical manufacturing of smart materials to achieve active 4D and multifunctional materials that are capable of sensing and responding to external stimuli with time. Examples of smart materials used in this study are shape-memory materials, piezoelectric and thermoelectric hybrids with main applications in smart textiles, artificial muscles and energy harvesting.

# Integrated Design of Magnetorheological Fluid and Energy Absorbers for Optimal System Performance

**Norman Wereley**

*Department of Aerospace Engineering, University of Maryland, College Park, MD 20742, United States of America*

*wereley@umd.edu*

**Abstract**

The ability to dissipate energy in vehicle systems, especially with the goal of protecting occupants from potentially injurious vibration, repetitive shock, crash and blast loads, is becoming a critical issue as the cumulative impact of these load spectra on chronic health and acute injury are becoming better understood. Typically, adaptive vibration and shock mitigation systems have utilized commercial-off-the-shelf magnetorheological fluids (MRFs) with specified properties as opposed to properties that are optimized for the particular application. The objective of this talk is to discuss what properties are optimal for a number of particular applications such as vibration isolation, shock load mitigation, or earthquake hazard mitigation. A number of key nondimensional parameters can be used to gain insight into how to define optimality for various applications including: Bingham number, Hedstrom number, Reynolds number, Mason number, dynamic range. Also, the trade-offs associated in designing an optimal MRF for a particular application are discussed. These trade-offs are illustrated using case studies from analytical, experimental and CFD studies.

# Design of Multifunctional Composites and Their Advanced Additive Manufacturing

**Daniel Therriault**

*Laboratory for Multiscale Mechanics (LM2), Polytechnique Montréal, C.P. 6079, succ. Centre-Ville, Montréal, QC, Canada*

*daniel.therriault@polymtl.ca*

**Abstract**

The manufacturing of smart composites composed of various functional components and built-in sensors is a serious technological challenge. Additive manufacturing (or 3D printing), known as a family of processes that joins materials to fabricate objects in a layer-by-layer fashion, is a powerful approach for the manufacturing of multifunctional systems. This talk presents an overview of our current research effort on the design of smart composites and the development of advanced additive manufacturing platforms featuring multi-material printing and robotic positioning systems. Examples of polymer-based composites featuring high mechanical stiffness and high temperature resistance, high electrical conductivity, piezoelectricity and printed structures with noise attenuation will be presented. The targeted applications are oriented toward the transportation, aerospace, microelectronics and biomedical fields.

# Acoustic Black Hole Structures and Their Applications in Vibration Damping and Noise Reduction

**Jinhao Qiu**

*State Key Laboratory of Mechanics and Control of Mechanical Structures, Nanjing University of Aeronautics and Astronautics, Nanjing 210016, China*

*qiu@nuaa.edu.cn*

**Abstract**

Acoustic black hole (ABH) effect utilizes the gradient variance of the structural configuration or material properties to realize the diminishing wave velocity in the structure. The wave velocity decreases to zero in an ideal scenario, resulting in zero reflection. The main method to realize the ABH structure is to adjust the structure through proper thickness tailoring in order to achieve energy capture in a certain area. It shows great advantages and potential application for flexural wave manipulation in thin-walled structure because of its high efficiency, broadband characteristic and flexible implementation. In this talk, the recent progress in modeling, analysis, implementation and measurement of ABH structures and their applications in wave manipulation, vibration damping and noise reduction is introduced. The topics in modeling, analysis, implementation and measurement of ABH structures include semi-analytical wavelet method for one-dimensional ABHs, FEM methods for two-dimensional ABHs and wave field visualization based on laser ultrasonic method. Applications of ABH structures include enhancement of energy harvesting, vibration damping based on energy focalization and ABH-based wide-band dynamic vibration absorber. Examples of cavity noise reduction using ABH panel is also introduced. Finally, the implementation of a one-dimensional ABH based on material stiffness gradation is demonstrated.

# Stiffness Modulation as a Method for Vibration Reduction

**Alexander Nowak***, **L. Flavio Campanile, and Alexander Hasse**

*Chemnitz University of Technology, Professorship Machine Elements and Product Development, Reichenhainer Str. 70, D 09126 Chemnitz, Germany*

**alexander.nowak@mb.tu-chemnitz.de*

**Abstract**

Among the different options for vibration reduction, methods involving periodic changes of structural stiffness gained relatively low attention by the scientific community up to now. Analysis of existing methods shows a heterogeneous picture, with essential differences concerning the time law of the stiffness variation, the physical device to implement the stiffness changes and, finally, the mechanism which leads to the vibration energy loss.

Most presented methods involve step changes at definite points of the vibration cycle and are demonstrated on systems with discrete degrees of freedom. In this paper, the case of continuous, periodic stiffness modulation of structures with distributed degrees of freedom is presented and analyzed. First, mass oscillator systems comparable to those in literature are introduced and the advantage of stiffness modulation is discussed. Thereafter, an experimental setup is presented to demonstrate how this benefit can be applied to continuous geometries. The case under study is a thin-walled beam with shape-adaptable cross-sections, which allows for adapting the beam's bending stiffness. Experimental and numerical tests are conducted on a thin steel plate provided with piezoelectric patches to realize the cross-section changes needed for stiffness adaptation.

Before dealing with stiffness modulation, the paper presents a survey of previous studies on vibration reduction based on stiffness variation. Existing methods are classified and compared to another, with focus on the physical energy-loss mechanism. The survey is complemented by numerical investigations on a representative choice of methods.

**Introduction**

Structures and machines are, in general, capable to experience vibrations due to the mutual interaction of elastic and inertia forces. In all cases in which vibrations are undesired and the inherent damping of a given system is insufficient to keep the decay time and the resonance amplitude below acceptable limits, vibration suppression techniques come into play, with the task of modifying the system's dynamics accordingly.

A multiplicity of approaches for vibration suppression are known from the literature, which can be classified in different ways. The probably most used classification subdivides the vibration suppression approaches into passive, active and semi-active methods. Passive methods are the ones which do not require actions by actuators or other controlled devices. Active methods interact with the vibrating system by controlled actuators which directly operate on the system's degrees of freedom. Semi-active methods involve controlled devices which act on the system's parameters.

Periodic changes of stiffness introduce a hysteretic loop in the force-displacement characteristic of the structure and can be therefore used as a basis for a semi-active method for vibration suppression. Seen in another way: properly chosen periodic stiffness changes redistribute the spectrum of the system's response over a broader range of frequencies, which is equivalent to a damping increase.

Published approaches suffer by limitations related to the particular device chosen for the stiffness adaptation. In (Xinghua, 2000) and (Ramaratnam and Jalili, 2006) only changes between two values of the stiffness are possible (switched stiffness); In (Yong et al., 2004) continuous stiffness changes are approximated by a system of two springs acting in parallel and connected by a friction interface. Besides the deviations with respect to the targeted behavior (stiffness continuously selectable by acting on the interface's normal force) the method has the drawback of involving friction, which makes discrimination between conventional and stiffness-related contribution to vibration suppression difficult.

A method for vibration suppression based on continuous stiffness adaptation (stiffness modulation) is still missing and will be presented in the paper.

**Methods/Model Formulations**

As a method for stiffness selection shape adaptation of the cross-section of thin walled beams is examined. The particular case under consideration is the beam showed in Figure 1. Under the action of the MFC piezo-patches the curvature of the beam's cross section can be increased, which results in a change of bending stiffness.

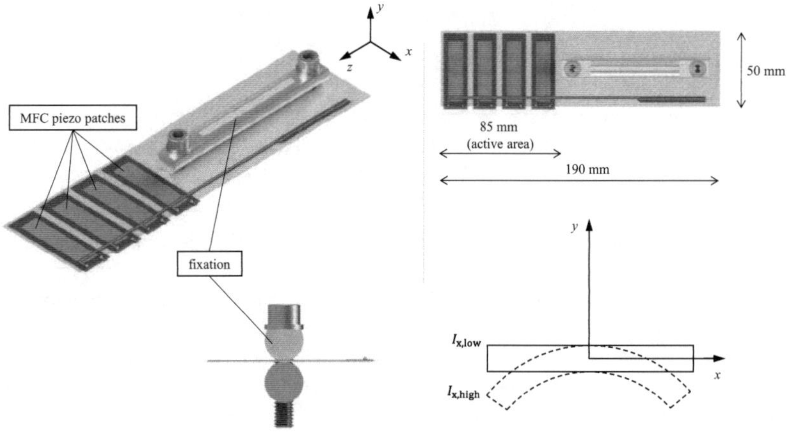


**Figure 1**. Experimental setup.

**Results and Discussions**

The obtained results confirm, in general, the positive expectations concerning the presented approach. In particular, the results of the free-vibration experiments show an increase of the time decay by up to 60%. FRF functions obtained by random and by swept sine excitation show a reduction of the highest peak by the same factor.

The effectiveness of the method depends on the time function used to modulate the stiffness. Smooth stiffness modulations show better results than square functions.

**References**


*Xinghua Y. (2000) Model and analysis of variable stiffness semiactive control system. Proceedings of the Twelfth World Conference on Earthquake Engineering, Auckland, New Zealand, Paper No. 1516.*

*Ramaratnam A and Jalili N (2006) A switched stiffness approach for structural vibration control: Theory and real-time implementation. J. Sound Vib. 291: 258–74.*

*Yong C, Zimcik D G, Wickramasinghe V K and Nitzsche F (2004) Development of the smart spring for active vibration control of helicopter blades. J. Intell. Mater. Syst. Struct. 15: 37–47.*

# On the Calculation and Control of Structural Intensities to Reduce Sound Transmission through Thin Walled Structures

**Michael Rose*** **and Alexander Kokott**

*German Aerospace Center (DLR), Lilienthalplatz 7, 38108 Brunswick, Germany*

**Michael.Rose@dlr.de*

**Abstract**

In lightweight and stiff structures like aircraft fuselages, the structure-born noise is often the dominating part in noise transmission. To efficiently use methods like active vibration reduction in order to attenuate this unwanted sound, suitable sensor information must be provided. In recent research, a new focus has been established on so-called structural intensities, a vector information on the transmitted sound energy per unit area as control target to efficiently block sound transmission within the structure.

This paper discusses methods to calculate and control structure intensities for panel like thin structures. For control purposes there is a need to at least estimate these quantities from real-time sensor data. In order to better understand the behavior of the structure with respect to sound transmission, offline calculations based on enhanced measurements e.g. from a laser scanning vibrometer are also discussed. In addition, the resulting power flow information can be used to adapt sensor and actuator positions.

**Introduction**

Vibrations from aircraft engines or other noise sources are often transmitted via structure-borne sound paths. The reduction of this transmitted noise can make a significant contribution to the attenuation of the radiated noise in cabins (aircraft, automobile). As input to control strategies like active vibration reduction, sensoric information is needed coming from devices like accelerometers or strain gauges. Often these methods achieve local reductions in certain areas of the structural path, but there still occurs major vibration energy transmission through the structure. Recent approaches (Sachau and Baschke, 2017) with respect to noise transmission show that the use of a suitable combination of lateral forces, torque components may attenuate the structure-borne noise transmission. The theory behind this approach is based on so-called structural intensity, a vector quantity describing structural born sound power flow per unit area (see Noiseux, 1970), as a control target for the noise reduction. Some recent research shows promising results in the case of beam and plate structures (Kokott, Haase and Monner 2019), especially in the case of inhomogeneous damping. To control structural intensities, the challenge of estimating these vector quantities from real-time sensoric input has to be addressed. This paper discusses the real-time determination as well as the offline calculation of structural intensities for thin walled panels. The former is limited by the amount of real-time sensors available, but inherently needed for control purposes. The latter can use enhanced measurements e.g. from a laser scanning vibrometer and delivers illuminating pictures of the power flow, which can be used to gain more insight into the structural behavior as well as to determine good sensor positions.

**Methods/Model Formulations**

Structural intensities depend on the elastic behavior of the underlying structure. They are expressed as suitable changes of curvature in combination with the underlying material constitutive equations. Derivatives up to the third order of deformation information must be determined, which may amplify the measurement noise in an unacceptable manner if not done properly. For the controller estimation, finite difference techniques are used in combination with suitable sensor arrays to provide the necessary input. Fig.1 shows a sample of a plate structure with disturbances (P) e.g. from a shaker, secondary control actuators (S) to reduce structural intensities estimated by sensor arrays at positions (M). The analysis includes methods of far and near field as well as different sensor principles. The barrier effect of the control systems can be shown by numerical simulations on a plate model, and challenges of the structure intensity measurement can be studied in experiments (measurement methods, sensor types, finite differences). The quality of the results directly depends on the amount of used sensors, their distances and the underlying finite difference technique. An optimal feedback control for the actuators was estimated using the disturbances as reference signal. To apply a linear filter as controller, the shear forces, moments, velocities and angular rates were used as error signals simultaneously.

For the offline calculation of structural intensity vector fields, suitable spline approximation methods have been investigated. In order to control the signal to noise ratio due to the evaluation of up to third order derivatives, smoothing techniques are inherently needed to filter flow information from the noisy data.

These methods require the solution of large saddle point problems. This implies, that even if large amount of sensor data is accessible in real-time, the current processing power will be far too slow to calculate the quantities in a control unit. But for offline calculation, the method delivers good results of the power flow.

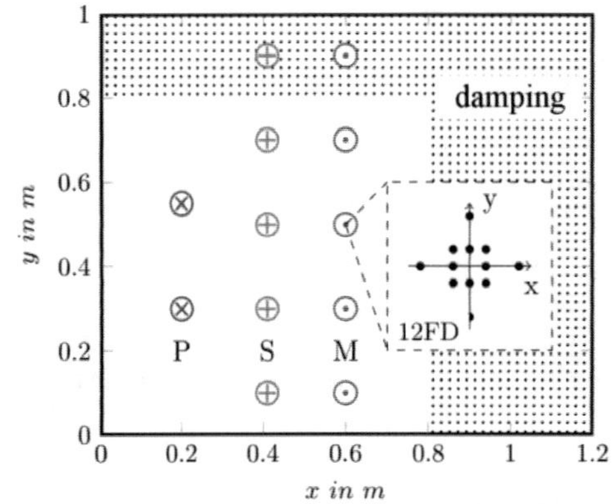


**Figure 1**. Sample plate structure to control structural intensities at (M) with disturbances at (P) and control actuators at (S).

**Results and Discussions**

The numerical experiments on plates reveal that the control of the structural intensities lead to a significantly higher global reduction of sound transmission (-30dB mean square reduction) in contrast to traditional velocity based methods (-15dB mean square reduction), which usually achieve primarily local attenuation. But there are severe issues with the signal to noise ratio. Complex finite difference schemes are yet not able to prevent noise amplification in contrast to simplified schemes, which only partly describe the full intensity vectors. More investigation and research is needed to improve these methods.

For the offline calculation, spline based approximation methods give reasonable results for the calculation of structural intensities, if appropriate smoothing is provided. The automatic determination of the tradeoff between approximation and smoothing is still been worked on, but the results are promising.

**References**

*Kokott A, Haase T and Monner HP (2019) Entwicklung einer Körperschallbarriere auf Basis der Strukturintensität: Teil 2. Deutsche Jahrestagung für Akustik 2019.*

*Noiseux DU (1970) Measurement of power flow in uniform beams and plates. The Journal of the Acoustical Society of America 47(1970): 238-247.*

*Sachau D, Baschke, M (2017) Adaptive support for aircraft panel testing: New method and its experimental verification on a beam structure. Journal of Sound and Vibration 394(2017): 31-45.*

# Vibration Control of MRE Isolator-embedded Smart Building using Genetic Algorithm

**Yang Yu[1], Yancheng Li[1*], Jianchun Li[1], Thuc N. Nguyen[1], Shaoqi Li[1], and Emre Erkmen[2]**

[1]*School of Civil and Environmental Engineering, University of Technology Sydney, Sydney 2007, Australia*

[2]*Department of Building Civil and Environmental Engineering, Concordia University, Montreal, QC, H3G 1M8, Canada*

[*]*yancheng.li@uts.edu.au*

**Abstract**

This study developed the adaptive genetic algorithm (GA) for vibration control of building structures subjected to ambient hazard excitations. An innovative smart building system was designed based on magnetorheological elastomer (MRE) isolators under each storey of the structure instead of being only installed beneath the entire structure. Such innovative system allows high authority semi-active control of storey responses by instantly changing the stiffness of the isolator, the control process of which can be considered as solving a global multi-objective optimization problem. Finally, a numerical investigation was conducted using a 5-storey international benchmark model under four benchmark earthquakes.

## Introduction

Seismic base isolation has been a widely adopted technique for earthquake protection of civil infrastructures. As the technology matures, new innovative designs of the base isolation systems become increasingly attractive to researchers, especially storey base isolation systems due to its design flexibility and better performance for seismic protection. Moreover, considering the unpredictable and diverse nature of earthquakes, the conventional base isolation systems have reached their limit due to their inherent passive nature which is incapable of adjusting their isolation frequencies according to the characteristics of the earthquakes.

Recently, with the advance of adaptive MRE base isolator, it provides an opportunity for the research and development on new adaptive base isolation systems. Accordingly, this study proposed an innovative semi-active storey isolation system. By inserting the MRE isolator between every floor, the storey isolation system is able to directly alter the stiffness of building structures by changing the applied current of the isolators of the corresponding floor and thus endows the building with the largest flexibility and high control authority. Therefore, such storey isolation system solves the problem that the base isolation system can hardly control the responses of higher levels above ground directly with the control force only applied on the ground floor. A five-storey benchmark building model has been utilized for numerical evaluation of the performance of this design. Comprehensive optimization work has been conducted in order to explore the optimal solution for structural parameters under passive control. Simulation results indicate that the seismic resistance performance of the building with optimal storey isolation system is significantly superior to the bare frame and building with passive base isolation system under different types of earthquakes.

## Methods

Highly adjustable adaptive MRE base isolators have been designed using the distinct characteristics and merits of MRE material (Li et al. 2013). Based on these innovative devices, a storey isolation system is constructed through incorporating the field-dependent MRE isolators under superstructure and between adjacent floors. Therefore, this storey isolation system is able to operate in two modes: 1) isolating mode, which means isolate every floor from the structure beneath it; 2) variable stiffness and damping mode, which means the stiffness and damping properties of the whole structure can be adjusted at each degree-of-freedom (DOF) by altering the applied current on the MRE isolator. To find out the optimal parameters of the system under passive control situation is of great significance in generating the guideline for structural design, which is also considered as solving an optimization problem. The key point for model parameter optimization is to set up a suitable fitness function. In this work, the peak inter-storey drift ratio, peak floor acceleration ratio, root mean square (RMS) inter-storey drift ratio, RMS floor acceleration ratio and peak floor relative displacement ratio are employed as indices for designing the fitness function, and the adaptive GA is adopted to identify optimal system parameters.

**Results and Discussions**

Figures 1, 2, 3 and 4 shows the comparisons of time-history inter-storey drift of 1st floor, peak inter-storey drift, peak acceleration and peak relative displacement between the proposed optimal system and other systems including bare building, storey isolated system without optimization, passive-(on and off) base-isolated system and base-isolated system with LQR controller, when the five-storey benchmark building is numerically excited by El-Centro earthquake. It can be seen from Figure 1 that obviously the proposed optimized storey isolation system has the best performance in mitigating the floor acceleration than other isolation systems. From Figure 2, the proposed system can well reduce the inter-storey drifts for all the floors compared to other systems. According to the results in Figure 3 and Figure 4, it indicates that the proposed optimal storey isolation system also outperforms other systems in terms of floor acceleration and relative displacement. Accordingly, the outcomes of this research provide a potential solution for the structural vibration suppression using the MRE device in the practical application.

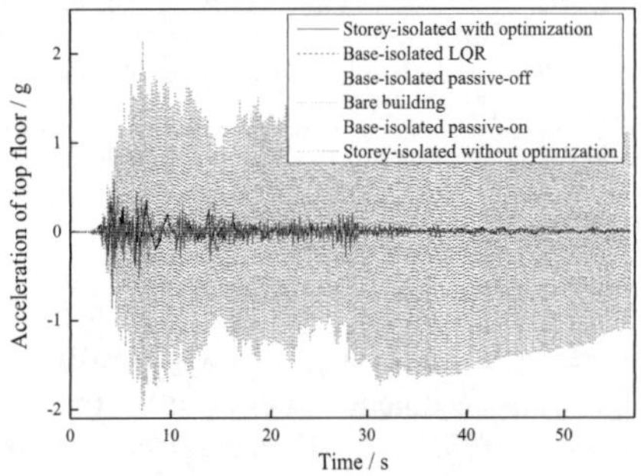

**Figure 1**. Time-history acceleration of top floor

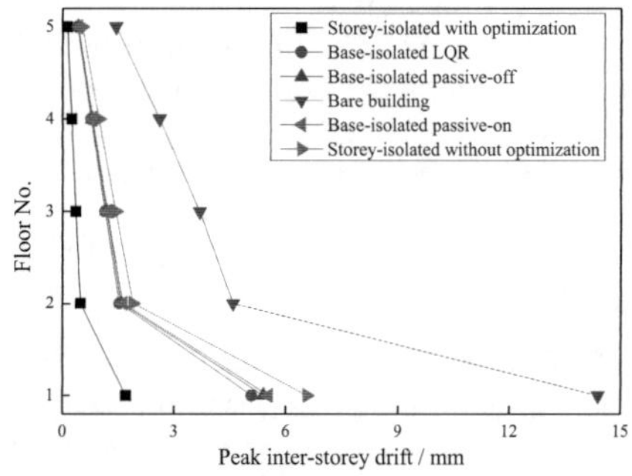

**Figure 2**. Peak inter-storey drifts

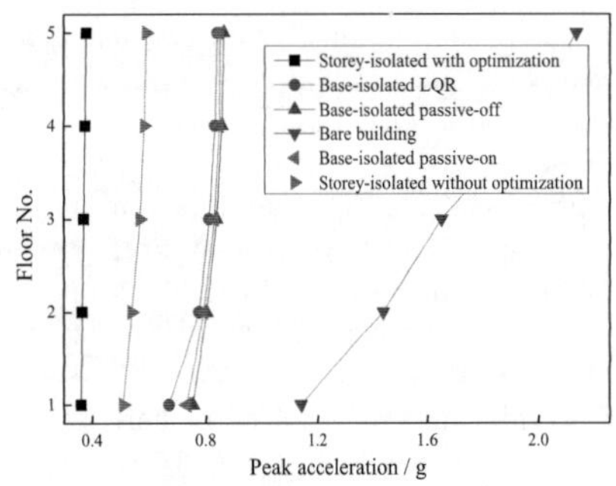

**Figure 3**. Peak acceleration

**Figure 4**. Peak relative displacement

**References**

*Li Y, Li J, Li W, et al (2013) A highly adjustable magnetorheological elastomer base isolator for applications of real-time adaptive control. Smart materials and structures 22(9): 129501.*

# Response Modification Factors for Friction Dampers as per the 2015 National Building Code of Canada

**Ali Naghshineh[1*], Fariborz M Tehrani[2**], Oscar Romero Galindo[3], and Ashutosh Bagchi[1]**

[1]*Department of Building, Civil and Environmental Engineering Concordia University, Montreal, QC, Canada, H3G 1M8*

[2]*Department of Civil & Geomatics Engineering, California State University, Fresno, California 93740-8030*

[3]*Quake Tek Inc., Montreal, QC, Canada, H4T 1T9*

[*]*ali.naghshineh@mail.concordia.ca,* [**]*ftehrani@csufresno.edu*

**Abstract**

This paper provides information on Response Modification Factors for Friction Dampers as per the 2015 National Building Code of Canada. The presented methodology considers diagonal braces for building frames with varied span lengths of 6m and 8m.

Keywords: seismic overstrength, response modification factor, performance-based design, concrete structures, friction dampers, moderately ductile.

**Introduction**

In the national building code of Canada (2015), the minimum earthquake lateral force is divided by Seismic Force Resisting System(s) (SFRS) reduction factor. This factor known as response modification factor which can be calculated by multiplying the overstrength factor (R0) and the ductility related force modification factor (Rd). The 2015 NBCC the overstrength factor (R0) and the ductility related force modification factor (Rd) are not considered for friction dampers. In this paper the SFRS reduction factors including the overstrength, ductility, and response modification factors of four-, ten-, and fourteen-story moderately ductile concrete frames with friction dampers, designed based on the NBCC 2015. The buildings are assumed to be located in Victoria, BC, in the western part of Canada with site class “C”.

**Nonlinear Static Pushover Analysis**

The 2015 National Building Code of Canada introduced the related force modification factors for different types of Seismic Force Resisting System(s) (SFRS) which is regardless of building height, span length and bracing configurations. In addition, these factors are not considered for friction dampers. In this research the method proposed by Miranda and Bertero (1994) in equations 1 and 2 was used for stiff soil to calculate the ductility reduction factor Rμ, where μ is ductility, T is the natural period of the structures and $\phi$ is a function of ductility, fundamental period and soil conditions. These results for Single Diagonal Braces (SBD) are presented in Table 1.

$$R_\mu = \frac{\mu - 1}{\varphi} + 1 \geq 1 \tag{1}$$

$$\phi = 1 + \frac{1}{10T - \mu T} - \frac{1}{2T} e^{-1.5|\ln(T) - 0.6|^2} \tag{2}$$

**Table 1**. Response modification factor for 4-, 8-, 14-Story

| Story | Bracing | Span Length | Overstrength factor $R_o$ | Ductility Reduction factor $R_\mu$ | Response modification factor |
|---|---|---|---|---|---|
| 4 | SBD | 6 | 3.06 | 2.60 | 8.01 |
| | | 8 | 2.38 | 2.55 | 5.93 |
| 8 | SBD | 6 | 2.20 | 3.51 | 7.74 |
| | | 8 | 1.91 | 2.97 | 5.64 |
| 14 | SBD | 6 | 1.47 | 4.43 | 6.52 |
| | | 8 | 1.41 | 3.92 | 5.53 |

## Inelastic Time History Analysis

In this part, 12 different ground motion records were selected and scaled based on the target response spectrum to assess the seismic performance of the friction dampers. Figure 1 presents the inter story drift ratio of different stories and span length .The sum of standard deviation and the mean values for 14-story building represent the inter-story drift ratio of 0.7% and 0.6% for 8m and 6m span, which is in the range of the "Life safety" performance of the structure defined in NBCC 2005 with the maximum drift at 36m height of the building. In 8-story these values are changed to 0.6%, and 0.5% for 8m and 6m span, and in 4-Story these values are 0.35% and 0.25% for 8m and 6m span.

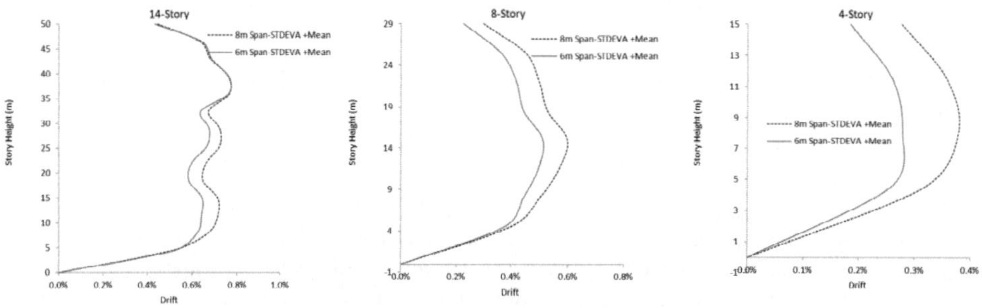


**Figure 1**. Matched Accelerograms based on the target response spectrum.

## Results and Discussions

Since the response modification factor for structures equipped with friction dampers varies with different structural parameters, such as span length and height of the buildings, the prescribed response modification factors can be recommended to NBCC 2015. The slip loads are calculated based on the lateral shear forces, shear deflection at each floor, lateral stiffness of braces alone as well as the lateral stiffness of existing frame and braces. Therefore, for simplicity and practical approach use by engineers, it can be suggested that the system may be modeled as a brace frame, and the axial loads in the braces should be equal or greater than the slip load of friction dampers. Moreover, the formation of the plastic hinges is reduced by almost 45% with friction dampers, which corresponds to the life safety performance.

## References

*Soong TT and Dargush GF (1998) Passive Energy Dissipation Systems for Structural Design and Retrofit. Multidisciplinary Center for Earthquake Engineering Research.*

*Lawsson RS, VanceV and Krawinkler H (1994) Nonlinear static pushover analysis- Why, When, and How Earthquake Engineering, Chicago, Vol.1, 283-292.*

*Miranda E, and Bertero VV (1994) Evaluation of Strength Reduction Factors for Earthquake-Resistant Design. Earthquake Spectra, 10(2), 357–379.*

# Aero-engine Vibration Transfer Path Analysis

**Seyed-Ehsan Mir-Haidari*** **and Kamran Behdinan**

*Department of Mechanical & Industrial Engineering, University of Toronto, 170 College St, Toronto, ON, Canada, M5S 3E3*

**ehsan.mirheidary@mail.utoronto.ca*

**Abstract**

Due to increasing demands for product comfort and safety, aero-engine manufacturers have been prominently invested in minimizing aero-engine vibrations. To address the vibration energy propagation issue through the aero-engine structure, Transfer Path Analysis (TPA) techniques are implemented to study vibration propagation in the structures and eventually propose design-based guideline to minimize vibration transfer. In this paper, using a reduced aero-engine model, the bond graph transfer path analysis is implemented to study vibration propagation and contribution in the structure to propose vibration reduction strategies and design guidelines for an aero-engine structure.

## Introduction

Due to advancement in technologies, the aero-engine manufacturing industry has placed prominent attention in minimizing ongoing operational costs. These innovative designs include efficient aero-engines with reduced overall structural mass, which has increased the sensitivity and exposure of the aero-engine to structural vibration caused by the aero-engine rotor system, which ultimately impacts the passenger's comfort due to higher than average noise and vibration. Due to these circumstances, greater attention has been placed on implementation Transfer Path Analysis (TPA) techniques in the early development stages to study root causes of vibration propagations in the aero-engines to ultimately proposing design modifications and guidelines (Van Der Seijs et al., 2016). Transfer Path analysis (TPA) designates the family of experimental and theoretical methodologies to investigate the transmission of mechanical vibration (Van Der Seijs et al., 2016) with the main goal of minimizing unwanted vibration energy propagation through a structure. A recently developed theoretical TPA methodology that is currently being researched on is the bond graph TPA method (Jalali Mashayekhi and Behdinan, 2018). Using the concept of transmissibility in conjunction with the bond graph methadology, a pure theoretical TPA method can be implemented. Due to the theoretical capabilities of this approach compared to its classical TPA counterparts, such as the Operational Path Analysis (de Klerk, 2009), the TPA bond graph model can significantly reduce aero-engine development costs by allowing evaluation of design-based modifications prior to any design prototyping and manufacturing. The goal of this investigation is to propose design methodology to minimize and attenuate vibration transfer in a reduced aero-engine model using the bond graph TPA technique by determining dominant vibration transfer paths. It can be used to determine vibration propagation in an aero-engine and further in its design. To the best of the author's knowledge, the proposed TPA methodology has never been used to determine vibration propagation in an aero-engine to propose relevant design guidelines.

## Methods/Model Formulations

The general formula that defined classical TPA is defined by Eq.(1):

$$\boldsymbol{y}(\omega) = \boldsymbol{H}(\omega).\boldsymbol{x}(\omega) \tag{1}$$

Where $\boldsymbol{x}$ is the input to the system, $\boldsymbol{y}$ is the system response, and $\boldsymbol{H}(\omega)$ is the structure-borne transfer function (de Klerk, 2009).To model an engineering system using bond graphs, the system is decomposed into subsystems connected to each other by ports (Karnopp et al., 2006), where the power that flows between the ports is called either flow or effort. The next step in performing bond graph TPA is to determine the state-space equation arranged into matrix formation (frequency domain) as shown by Eq. (2-3):

$$\vec{\mathbf{X}} = (\mathbf{sI} - \mathbf{A})^{-1}\mathbf{B}\vec{\mathbf{U}} \tag{2}$$

$$\vec{\mathbf{Y}} = [\mathbf{C}(\mathbf{sI} - \mathbf{A})^{-1}\mathbf{B} + \mathbf{D}]\vec{\mathbf{U}} \tag{3}$$

where $\vec{X}$ is the state variables, $\vec{Y}$ is the system output, $\vec{U}$ is system input, **A** is the state matrix, **B** is the input matrix, **C** is the output matrix, and **D** is the direct transmission matrix. Using global transmissibility concept and mathematical manipulations, the transmissibility at various frequency ranges can be defined by Eq. (4) (Jalali Mashayekhi and Behdinan, 2018):

$$TG_{ij} = \frac{[C^j(sI - A)^{-1}B + D^j]}{[C^i(sI - A)^{-1}B + D^i]} \tag{4}$$

**Methods/Model Formulations**

A reduced aero-engine model adapted from short range aero-engine is depicted in Fig.1. In order to perform bond graph TPA, the bond graph model corresponding to Fig. 1 is developed and is illustrated in Fig.2 using the bond graph methodology. Using the developed bond graph, the state-space equation can be readily, which allows the determination of the global transmissibility between various inertia elements. The state-space equation is illustrated in Fig. 3.

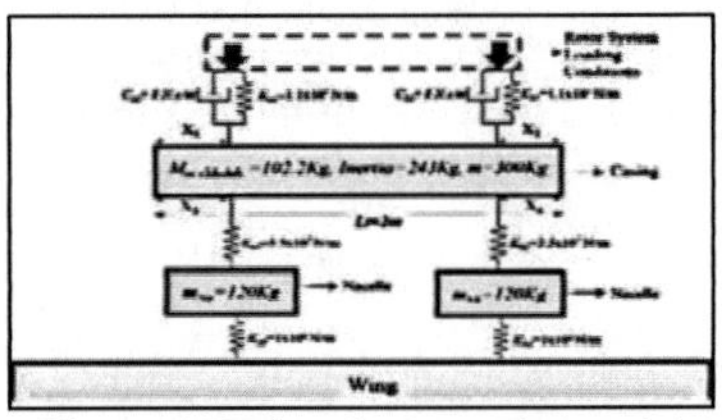


**Figure 1**. Reduced aero-engine model

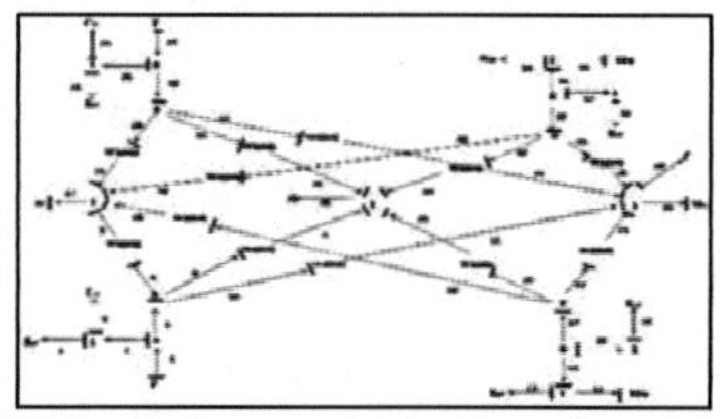

**Figure 2.** Aero-engine bond graph model

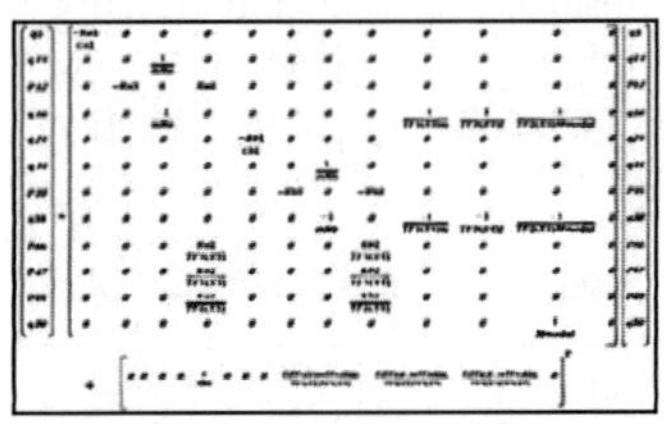

**Figure 3**. State-space equation ($\vec{\dot{x}} = \mathbf{A}\vec{x} + \mathbf{B}\vec{u}$)

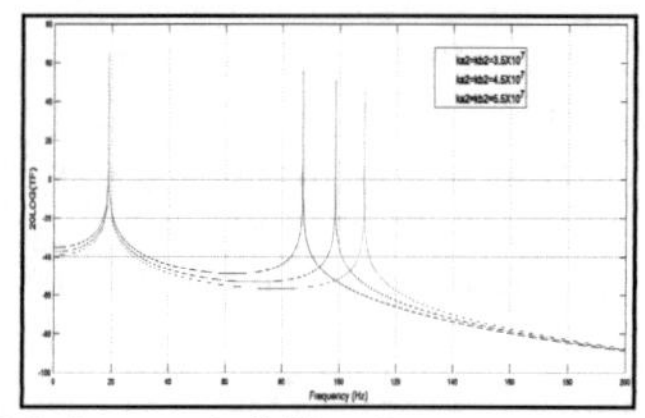

**Figure 4**. Global transmissibility: Rotor system and the nacelle

**Results and Discussions**

To demonstrate the theoretical capabilities of the proposed bond graph TPA methodology for implementation on an aero-engine model, the physical parameter connecting the aero-engine to the nacelle (Ka2 and Kb2 stiffness values) is analyzed as a possible design parameter to determine the effects of the vibration contribution calculated as the global transmissibility taking place form the rotor (source) to the nacelle (receiver). The stiffness values that are implemented to analysis the structural connection effects are Ka2=Kb2= 4.5x107N/m and 5.5x107N/m. As observed form Fig. 4, it is evident that by increasing the stiffness values corresponding to Ka2 and Kb2, the vibration propagation in the aero-engine model is minimized due to reduction in the transmissibility peaks magnitude. In practicality, this can be achieved by proper selection of structural materials for the aero-engine structural connections.

**References**


*de Klerk D (2009) Application of Operational Transfer Path Analysis on a Classic Car. Daga 2009: 4–7.*

*Jalali Mashayekhi M and Behdinan K (2018) Analytical transmissibility based transfer path analysis for multi-energy-domain systems using bond graphs. JVC/Journal of Vibration and Control 24(13): 2927–2937. DOI: 10.1177/1077546317696362.*

*Karnopp DC, Margolis DL and Rosenberg RC (2006) SYSTEM DYNAMICS. Fourth. John Wiley & Sons.*

*Van Der Seijs M V., De Klerk D and Rixen DJ (2016) General framework for transfer path analysis: History, theory and classification of techniques. Mechanical Systems and Signal Processing 68–69. Elsevier: 217–244. DOI: 10.1016/j.ymssp.2015.08.004.*

# Optimal Magnetorheological Damper for Two-Wheeled Vehicle

**Manjeet Keshav*, Eashan Dhawade, and Sujatha Chandramohan**

*Department of Mechanical Engineering, Indian Institute of Technology Madras, Chennai-600036, India.*

*manjeetkshv@gmail.com

**Abstract**

Magnetorheological (MR) fluid technology has gained significant development in effectively isolating undesirable vibrations by use of MR fluids acting as a dissipating medium in a damper. The current work aims at optimally designing an MR damper for a commercial 110 cc two-wheeled vehicle; this damper would replace the conventional passive damper. The geometric dimension of the valve is designed in such a way that it should match the performance of the passive damper in field-off state and includes dimensional constraints from the geometry of the vehicle. After the dimensions of the damper are decided on, valve geometry is optimized in MATLAB environment for multiple objective functions using an analytical method of solving a magnetic circuit. Further, the optimized MR damper is implemented in a 5-DOF half-car vibration model developed by considering bounce and pitch motions of the vehicle and the response is calculated at the rider's seat by incorporating the MR damper at the rear end and a passive damper at the front end. The response is then compared with that of the vehicle model with conventional passive dampers at both ends.

**Introduction**

Ride comfort is considered as an important factor in the design of an automobile. Application of magnetorheological (MR) damper has shown outstanding capability in attenuating automobile vibrations; this has led to the use of MR dampers in automobiles like sports cars, military vehicles and also in some helicopters. However, minimal studies have been conducted for 2-wheeled vehicle applications. Technical and geometric data for the rear passive damper which is to be replaced is given in Table (1). The MR damper has to match this minimum performance and geometric requirements. The material of the valve inside the damper has been chosen as Silicon steel and the MR fluid is MRF-132DG (Sung and Choi, 2008). Following the data of the rear passive damper in Table (1), requirements of the MR damper have been enlisted in Table (2).

**Table 1**. Technical and geometric data of passive damper

| Parameter | Value |
|---|---|
| Damping coefficient (tension) | 842.2 Ns/m |
| Damping coefficient (compression) | 233 Ns/m |
| Outer diameter | 40 mm |
| Damper length | 315 mm |

**Table 2.** Requirements of MR damper.

| Parameter | Value |
|---|---|
| Damping coefficients (tension) | 800 Ns/m |
| Damping coefficient (compression) | 200 Ns/m |
| Length of piston valve | 60 mm |
| Diameter of piston valve | 30 mm |

**Optimization**

For the geometric optimization of the MR valve inside the damper, the analytical method proposed by Nguyen et al. (2009) has been used where a single design variable, i.e., piston housing thickness, th, is considered to optimally design the whole geometry. In the present optimization problem, the objective is to maximize damping force and dynamic range, and minimize inductive time constant with equal weightage given to all objective functions. The optimal geometry of the MR valve is given in Table (3).

**Implementation in 5-DOF Vehicle Model**

The vibration model is excited by two bumps modelled by sinusoidal lobes as shown in Fig. 1 where l = 0.6 m and a = 0.102 m. The optimized MR damper is placed in the rear side of the two-wheeled vehicle and the front side contains the conventional passive damper (Fig. 2). For controlling the damper force, Skyhook control strategy has been used. For the optimized damper, cmin is 200 Ns/m at 0.04 A of current and cmax is 810 Ns/m at 0.7 A of current.

**Table 3.** Optimal geometry of piston MR valve.

| Dimension | Value (mm) |
|---|---|
| Piston outer radius, R | 15 |
| Piston legnth, L | 60 |
| Housing thickness, $t_h$ | 4 |
| Coil width, $w_c$ | 3 |
| Core radius, $R_c$ | 7 |
| Pole length, $L_p$ | 24 |

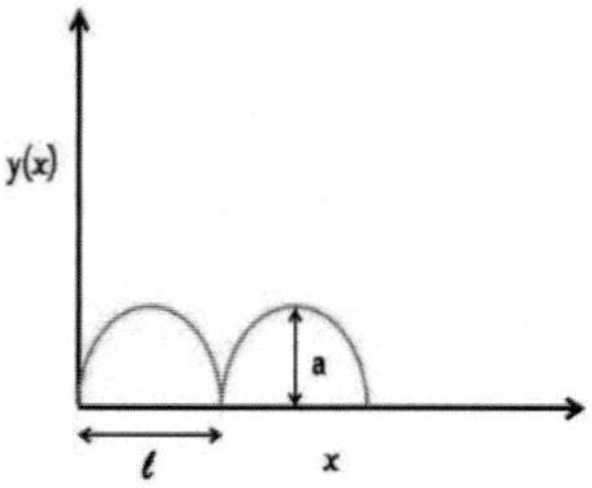


**Figure 1.** Profile for road input.

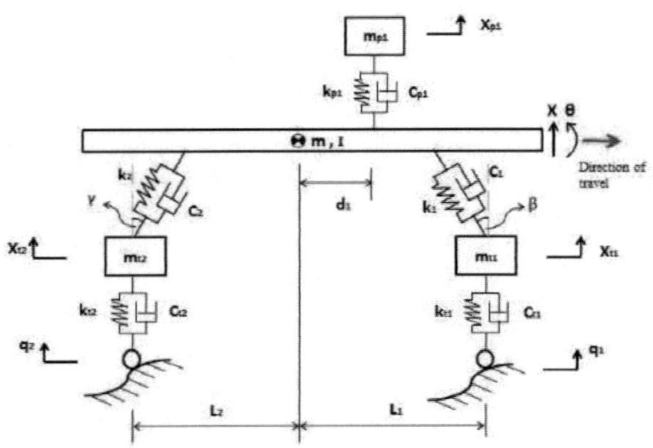


**Figure 2.** 5-DOF two wheeld vehicle.

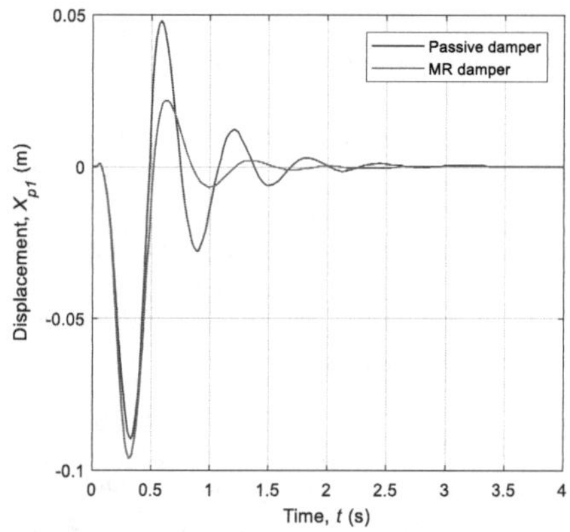


**Figure 3.** Displacement response at driver's seat.

## Results and Discussion

Figure 3 compares the displacement response of the passive and optimal MR dampers. The response wit MR damper showed 54.8% improvements in the displacement with peak value for MR damper as 0.0217 m as compared to 0.048 m for passive damper.

## Conclusion

The MR damper is designed considering the geometric dimensions of the passive damper. Thus, only the passive damper needs to be replaced by the MR damper without making any changes to the rest of the structure of the suspension system. The 5-DOF model simulated the response of the vehicle fitted with rear semi-active suspension system fairly accurately. In the event of failure or absence of power supply, the MR damper will work as a passive damper with its viscous damping characteristics similar to the damping characteristics of passive damper. Thus, the malfunctioning of electronic components will not lead to the failure of the rear suspension system.

## References


*Nguyen QH, Choi SB, Lee YS and Han MS (2009) An analytical method for optimal design of MR valve structures. Smart Materials and Structures, 18 095032 (12pp).*

*Sung KG and Choi SB (2008) Effect of an electromagnetically optimized magnetorheological damper on vehicle suspension control performance. Proc. IMechE Part D: J. Automobile Engineering, 222 2307-19.*

# Intelligent Resonance Identification and Vibration Suppression for Industrial Robots

**Michael Newman, Ehsan Jalayeri, and Matt Khoshdarregi***

*Intelligent Digital Manufacturing Laboratory (IDML), Department of Mechanical Engineering University of Manitoba, 75A Chancellors Circle, Winnipeg, Manitoba R3T 5V6 Canada*

**M.khoshdarregi@umanitoba.ca*

**Abstract**

Due to their long kinematic chain, industrial robotic arms exhibit pose-dependent low-frequency structural modes at their end-effector. In high speed positioning motions, the resonance modes of the arm are excited, leading to significant residual vibrations. The frequency of the vibrations depends on the robot's pose and the carried payload. This paper presents a trainable framework based on an artificial neural network (ANN) which predicts the frequency of structural vibrations given the robot's joint angles and the carried payload. The experimentally trained vibration model is then used to design a pose-dependent Input Shaper to avoid the excitation of structural modes in rapid positioning motions. The developed techniques have been experimentally implemented on a 6-DOF Staubli RX90 robot with an in-house developed open-architecture controller. Preliminary results demonstrate a proof of concept.

**Introduction**

Structural vibrations of a robotic arm stem primarily from the flexibilities in the gearbox of the first two joints, i.e. base and shoulder (Kim and Croft, 2018). Input shaping technique (IST) is a feedforward approach that modifies the commanded trajectories to avoid excitation of a structure's natural frequencies (Singhose, 2009). Due to the pose-dependency of a robot's equivalent inertia and stiffness, IST can only be applied for a known configuration and payload. Chang (Chang and Park, 2005) developed an empirical pose-dependent model that schedules the vibration frequency of a robot based of a calculated swing length from the base to the end effector. This paper aims at developing a trainable ANN framework that correlates the vibration frequency directly to the joint angles and payload. Such a framework can later be used along with machine learning techniques to allow a robot to intelligently identify its dynamic behavior based on a set of test motions. The self-identified model can then be used to tune an input shaper for any given pose and payload to avoid residual vibrations.

**Methods/Model Formulations**

For a simple structural mode with a natural frequency of $\omega_n$ and a damping ratio of $\zeta$, a Zero-Vibration-Derivative (ZVD) input shaper comprises three impulses with the following amplitudes

$$\begin{bmatrix} A_1 \\ A_2 \\ A_3 \end{bmatrix} = \frac{1}{K^2+2K+1}\begin{bmatrix} 1 \\ 2K \\ K^2 \end{bmatrix} \rightarrow K = e^{-\pi\zeta/\sqrt{1-\zeta^2}} \tag{1}$$

Each impulse is shifted by $T_d/2$ with respect to the previous impulse, where $T_d$ is the period of vibrations. It can be shown that if this input shaper is mathematically convolved with any custom trajectory, the so-called shaped command will not cause any residual vibrations. For multi-mode systems, input shapers can be designed individually for each mode and convolved together to avoid all of the considered modes.

In order to experimentally extract the natural frequencies of a robot, the common practice is to conduct manual hammer tests at the end effector for different poses and payloads (Glogowski et al., 2016). Impact hammer testing is a time consuming and labor intensive process. This paper proposes an autonomous framework that allows a robot to excite its natural frequencies at different poses by sending a high jerk input command to the joints, collect vibration data using accelerometers, and identify the dominant structural modes using Fast Fourier Transform (FFT). A trainable framework based on ANN allows the robot to autonomously repeat these tests over a set of grid points in the workspace and train the vibration model.

In order to implement the proposed framework, an open-architecture control system has been developed for the first three joints of a Staubli RX90 industrial robot. Each joint has been retrofitted with an ABB Microflex Analog servo drive operating in torque (current) mode. The joints are controlled by P-PI controllers implemented via MATLAB Simulink and Quanser data acquisition board. The developed system allows the user to implement custom control laws and generate any desired trajectory commands for the joints.

### Results and Discussion

For the preliminary verifications of the proposed resonance identification and vibration suppression techniques, a flexible oscillator in the form of an aluminum link (1"x1/8"x12") has been attached to the end effector of the robot. A motion command with an amplitude of 10° has been interpolated with a trapezoidal velocity profile and sent to Joint 1. The flexible link has a natural frequency of $\omega_n = 24$ Hz, which is relatively close to the robot's structural frequencies (14 Hz - 18 Hz). The link's natural frequency can be varied by adding tip-weights to simulate the varying natural frequency of the robot as a function of pose. Given the identified natural frequency, an input shaper is designed using Eq. (1) for each tip-mass. Figure 1 shows sample experimental results for the identification and suppression of the oscillator's vibrations in high speed positioning motions of the robot. The implementation of the proposed techniques for the suppression of the robot's own pose-dependent vibration modes is currently being investigated.

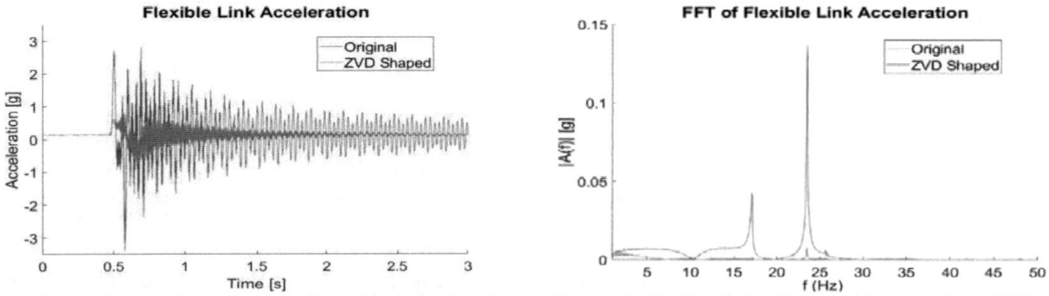


**Figure 1**. Sample experimental results for multi-mode vibration avoidance of a flexible link with natural frequencies at 17 Hz and 24 Hz.

### References


*Chang PH and Park HS (2005) Time-varying input shaping technique applied to vibration reduction of an industrial robot. Control Engineering Practice 13(1): 121–130.*

*Glogowski P, Rieger M, Sun J, et al. (2016) Natural Frequency Analysis in the Workspace of a Six-Axis Industrial Robot Using Design of Experiments. Advanced Materials Research 1140: 345–352.*

*Kim J and Croft EA (2018) Preshaping input trajectories of industrial robots for vibration suppression. Robotic and Computer Integrated Manufacturing 54: 35–44.*

*Singhose W (2009) Command shaping for flexible systems: A review of the first 50 years. International Journal of Precision Engineering and Manufacturing 10(4): 153–168.*

# Dynamic Analysis and Design Optimization of a Light-weight Magnetorheological Elastomer-based Adaptive Vibration Absorber

**Mostafa Asadi Khanouki***, **Ramin Sedaghati, and Masoud Hemmatian**

*Department of Mechanical, Industrial and Aerospace Engineering,*

*Concordia University, Montreal, QC, Canada, H3G 1M8*

*mostafa.asadi@concordia.ca

**Abstract**

The present study aims to investigate the dynamic performance of a novel magnetorheological elastomer based adaptive vibration absorber (MRE-based AVA). The proposed MRE-based AVA consists of a light-weight sandwich beam treated with an MRE core layer and two electromagnets at both free ends. The MRE layer provides the opportunity to vary the stiffness of the absorber by applying the required magnetic flux density using the electromagnets. The electromagnets are designed to provide the demanded magnetic flux density while also acting as the mass of the absorber. The finite element (FE) model of the proposed MRE-AVA has been developed to evaluate the frequency range of the absorber under varying magnetic field intensity. Finally, a multidisciplinary design optimization strategy has been formulated to maximize the dynamic range of the proposed MRE-based AVA while respecting the constraints of weight, electromagnets design, and sandwich beam geometry and deflection.

**Introduction**

Vibration and noise control has been an inseparable part of the engineering design and research studies since the onset of 20th century (Frahm, 1911). Several designs have been proposed in the literature for vibration absorbers based on MR materials. Lerner et al. (2008) have designed three types of MRE-based vibration absorbers working in shear, squeeze, and longitudinal modes among which the squeeze mode absorber exhibited the widest frequency range of 57-347 Hz under applied magnetic field of 0kA/m to 20 kA/m. There are also other research works that investigated the application of MREs in adaptive vibration absorbers (Sun et al., 2015; Yang et al., 2015; Xin et al., 2016); however, limited works have been conducted on the optimization of absorbers to achieve the best performance.

In the present work, a three-layer sandwich beam with thin aluminum elastic layers on top and bottom faces and MRE as the core layer is modeled using FE method. Three potential designs are considered for the electromagnets and a magnetic analysis is performed using FEMM software to evaluate their ability in providing the required magnetic flux density. FE model of the sandwich beam treated with MRE in the core layer, which is activated only in specified regions under generated field of electromagnets, is established and frequency range of the absorber is obtained for each of the considered designs. Finally, a design optimization problem is formulated on geometrical parameters of the beam and location of the electromagnet to maximize the frequency bandwidth of the absorber.

**Finite Element Modeling of MRE-Based ATVA**

Figure 1 shows the schematic of the sandwich beam on the left and the assembled absorber structure on the right. One H-shape electromagnet is attached at the both ends of the beam that has the role of generating magnetic field over the specified parts of the MRE layer as well as acting as oscillating mass of the absorber. Figure 2 shows the electromagnet design along with magnetic analysis of the considered design in FEMM software. As can be seen in the Figure 2, the H-shape electromagnet can provide flux density of 1 T on the MRE core layer.

**Results and Discussion**

The FE problem is solved considering a sandwich beam with 20 cm length. The 3 mm core layer is made of MRE with 40% volume fraction of CIP and silicone rubber matrix (Asadi Khanouki et al., 2019). Density of the MRE is 3500 kg/m3.

The aluminum face layers have a thickness of 1 mm, density of 2500 kg/m3, and elastic modulus of 70 GPa. Table 1 presents the results of the FE problem. Using the H-shape electromagnet, magnetic flux density of 1 T could be generated which can change the modulus of the MRE from 100 kPa at zero field to 2991 kPa. Consequently, the fundamental frequency of the absorber varies from 37.45 Hz to 58.04 Hz. Finally, the design optimization problem is formulated and solved to maximize the operating frequency range of the MRE-based absorber while respecting constraints on beam static deflection and total weight of the structure.

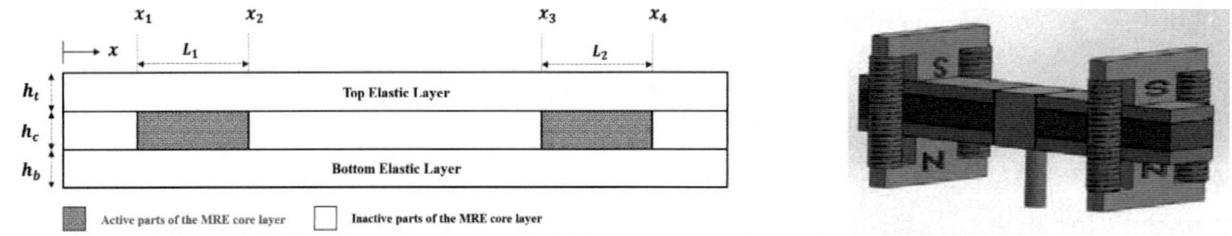


**Figure 1**. Left: Schematic of the sandwich beam structure. Right: Assembled absorber structure

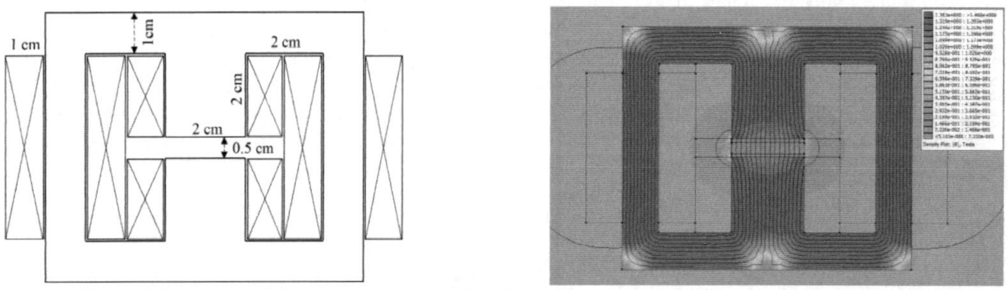


**Figure 2**. Left: Dimensions of the electromagnet, Right: Magnetic analysis of the electromagnet in FEMM software

**Table 1**. Results of the FEM problem regarding the absorber's frequency range and beam's tip deflection

| Position of MRE patches subjected to magnetic field | Electromagnet Mass (kg) | Magnetic Field (T) | Tip Deflection (cm) | Absorber's Fundamental Natural Frequency Range (Hz) |
|---|---|---|---|---|
| [x=1 cm to x=3 cm] symmetric | 1.22 | 1.06 | 1.45 | 37.45 -58.04 |

## References

*Asadi Khanouki M, Sedaghati R and Hemmatian M. (2019) Characterization of Magneto-Mechanical Properties and Quasi-Static Physical Modelling of MR Elastomers. AIAA Scitech 2019 Forum. 1114.*

*Frahm H. (1911) Device for damping vibrations of bodies. Google Patents.*

*Lerner AA and Cunefare KA. (2008) Performance of MRE-based vibration absorbers. Journal of Intelligent Material Systems and Structures 19: 551-563.*

*Sun S, Deng H, Yang J, et al. (2015) Performance evaluation and comparison of magnetorheological elastomer absorbers working in shear and squeeze modes. Journal of Intelligent Material Systems and Structures 26: 1757-1763.*

*Xin F-L, Bai X-X and Qian L-J. (2016) Principle, modeling, and control of a magnetorheological elastomer dynamic vibration absorber for powertrain mount systems of automobiles. Journal of Intelligent Material Systems and Structures: 1045389X16672731.*

*Yang C, Fu J, Yu M, et al. (2015) A new magnetorheological elastomer isolator in shear–compression mixed mode. Journal of Intelligent Material Systems and Structures 26: 1290-1300.*

# The Effectiveness of Distributed Tuned Mass Damper Integrated with Adaptive MR Damper in Building

**A.Bagchi[*] and A. Torkaman Rashid[**]**

*Department of Building, Civil & Environmental Engineering, Concordia University, Montreal, QC, Canada*

[*] *ashutosh.bagchi@concordia.ca*

[**] *a_torka@encs.concordia.ca*

**Abstract**

Energy dissipation devices are utilized in buildings to improve their resilience and safety against dynamic forces such as seismic loads. Since the most commonly used passive dampers have operational limitations due to lack of control on their performance, in recent years, engineers tend to employ adaptive hybrid dampers which could be adjusted according to the seismic demands. In this research, seismic behavior of a fifteen story building equipped with 3 tuned mass dampers (TMD) is investigated and compared with the seismic response of the same building in which 3 magnetorheological dampers (MR) are paired with the TMDs to operate concurrently with them as hybrid energy dissipating system. The MR dampers are commanded by a linear quadratic regression (LQR) control method which adjust the dampers resistance force according to the seismic displacement and velocity of the structure. This building is subjected to 5 far field earthquake records and the results display that the hybrid system can perform considerably better than passive TMDs improving the seismic demands of the structure.

## Introduction

Earthquake is a natural phenomenon that cannot be predicted and anticipated. In order to prevent the damages caused by earthquake, civil engineers have studied many ways to make the design of the structure resistant to earthquake. During recent years, many methods have been discovered to mitigate the vibration of the structure due to ground motions. Some of the methods have been implemented in the design of the structure to make them more resilient to lateral forces applied. Furthermore, many types of devices have been developed to dissipate the energy of the applied earthquake to the structure. The utilization of a mass for mitigating of the structure vibration have been used for many years. The system that would use a mass with a stiffness for the reduction of ground motion effects on the structure is called Tuned Mass Damper (TMD). This system consist of a mass with a damping and stiffness. Many studies have been conducted for designing the optimum characteristics of a mass damper. Primarily, structure with a tuned mass damper located at the top floor have been studied. They have been designed in such a way that could have the best response due to the fundamental vibration mod of the structure. Meanwhile, many optimization methods has been used for determining the efficient parameters of TMD. During the first decade of 20th century, the effects of Multiple Tuned Mass Dampers (MTMD) for a one degree of freedom structure have been studied. It has been determined that a structure with MTMD tuned to distributed frequencies has better response to a single TMD at the top of the structure. Furthermore, the modelling and design of single active and semi-active Tuned Mass Dampers have been studied (Li and Zhu, 2007). It has been proven that the utilization of a semi-active TMD would be more beneficial in the aspect of saving energy and the resilience of the structure.

## Design of Tuned Mass Damper

For the design of the Tuned Mass Damper, it is needed to determine the amount for the parameters of a TMD, which are the mass, the stiffness, and the damping coefficient. The mass of the damper is usually chosen about the 5 percent of the mass of the whole structure. The stiffness of the TMD is chosen such that the TMD would be tuned to the natural frequency of the structure. The following formula is for determining the parameters of tuned mass damper (Berberian et al., 1992).

$$\frac{\omega_d^2}{\omega_s^2} = \frac{2+\mu}{2(1+\mu)^2}$$

$$\mu = \frac{m_d}{M_s}, \qquad \omega_s^2 = \frac{K_s}{M_s}, \qquad \omega_d^2 = \frac{k_d}{m_d} \tag{1}$$

Here, $M_s$ and $K_s$ are the Mass and Stiffness of the structure, respectively. And$m_d$, $k_d$ are the mass and stiffness of the tuned mass damper. By determining the mass of the damper to be 2 percent of the mass of the structure and tuning the damper to the principal frequency of the structure, the stiffness of the damper could be determined using equation (1).

The following equation is used for determination of the non-dimensional damping coefficient (McNamara, 1977):

$$\gamma = \sqrt{\frac{\mu(4+3\mu)}{8(1+\mu)(2+\mu)}} \tag{2}$$

Finally, the value of the damping coefficient could be calculated using equation (3).

$$c_d = 2\gamma\omega_d m_d \tag{3}$$

**Equation of Motion**

The equation of motion for a multi-degree of freedom structure is shown in equation (4).

$$M\ddot{x} + C_s\dot{x} + Kx = \Gamma f - M\Lambda\ddot{x}_g \tag{4}$$

The equation of the state space is given by

$$\begin{cases} \dot{x} = Ax + Bu \\ y = Cx + Du \end{cases} \tag{5}$$

**Response of a Multi-degree of Freedom Structure**

It is observed that by equipping the structure with the MTMD energy dissipating system the average maximum seismic displacement, velocity, and acceleration is decreased by about 27, 29, and 14 percent respectively. Meanwhile the induced vibration energy of the top floor is reduced by about 76 percent. The analytical results illustrate that by installing MR dampers paired with each TMD, the structure would have a notable effective energy dissipating system that mitigate the seismic demands of the structure including the average of maximum displacement, maximum velocity, maximum acceleration, and vibration energy of the top floor by about 67, 61, 35, and 97 percent respectively.

**References**


*Berberian M, Qorashi M, Jackson J, et al. (1992) The Rudbar-Tarom earthquake of 20 June 1990 in NW Persia: preliminary field and seismological observations, and its tectonic significance. Bulletin of the seismological society of America 82: 1726-1755.*

*Li C and Zhu B. (2007) Investigation of response of systems with active multiple tuned mass dampers. Structural Control and Health Monitoring: The Official Journal of the International Association for Structural Control and Monitoring and of the European Association for the Control of Structures 14: 1138-1154.*

*McNamara RJ. (1977) Tuned mass dampers for buildings. Journal of the Structural Division 103: 1785-1798.*

# Effect of Parametric Uncertainties on Vibration Mitigation of a Rod with Interconnected Piezoelectric Patches

**Marcelo A. Trindade[1*], Boris Lossouarn[2], and Jean-François Deü[2]**

[1]*Department of Mechanical Engineering, São Carlos School of Engineering, University of São Paulo, Av. Trabalhador São Carlense, 400, São Carlos, SP, 13566-590, Brazil*

[2]*Laboratoire de Mécanique des Structures et des Systèmes Couplés (LMSSC), Conservatoire national des arts et métiers (Cnam), 292 Rue Saint-Martin, 75003 Paris, France*

**trindade@sc.usp.br*

**Abstract**

This work aims to present an analysis of the effect of parametric uncertainties on the vibration control performance of a rod with periodically distributed piezoelectric patches that may be either independently connected to electrical shunt circuits or interconnected through an electrical line of inductors. In both cases, the capacitance of piezoelectric patches are considered as stochastic parameters following a known probability density function distribution. Then, Monte Carlo simulations are performed to evaluate mean values and confidence intervals of the frequency response functions to assess the robustness of each solution and to compare different solutions in terms of nominal and robust performances. The effect of other electrical components, such as circuits resistances, on the robustness is also analyzed.

## Introduction

Piezoelectric materials have been widely used as distributed sensors and actuators to control structural vibrations of flexible structures. They are normally bonded to a host structure and connected to an electrical circuit. Therefore, they provide an electromechanical coupling to the host structure that, depending on electrical connections, may be used for several purposes such as passive, semi-passive, active or active-passive vibration control (Santos and Trindade, 2011).

In the case of passive vibration control applications, most studies focused on shunt circuits that connect the two electrodes of each piezoelectric patch. The most common circuits are resistive and resonant (resistive-inductive) ones but several research works proposed more sophisticated circuits using, for instance, switch or negative capacitance circuits. More recently, Lossouarn et al. (2015) considered the interconnection of a network of piezoelectric patches through resonant electrical circuits in a way that good multimodal vibration attenuation performance may be obtained with much smaller inductance values, compared to resonant shunt circuits.

However, as for the shunted case, the vibration control performance is dependent on the adequate tuning between electrical circuit components and structure's mechanical properties. Thus, variabilities and/or uncertainties on material properties, boundary conditions, and environmental conditions may have a major effect on reducing the expected or predicted performance of such devices. Nevertheless, few studies attempted to analyse the effect of parametric uncertainties on passive vibration control using shunted piezoelectric patches (Santos and Trindade, 2011; Andreaus and Porfiri, 2007).

## Problem Description

A free-free AU4G duraluminum rod in longitudinal vibration with length 1000 mm, width 20 mm and thickness 20 mm is considered. The rod has mass density 2780 kg $m^{-3}$ and Young modulus 73.9 GPa. 20 pairs of PIC151 piezoelectric patches are bonded to the rod. Each patch has the following dimensions: length 30 mm, width 20 mm and thickness 0.5 mm. The piezoelectric patches are considered to be equal with mass density 7800 kg $m^{-3}$, Young modulus 66.7 GPa, piezoelectric charge constant -210 pC $N^{-1}$ and dielectric permittivity 21.2 nF $m^{-1}$.

The rod with periodically bonded piezoelectric patches was modeled using the transfer matrix approach. For that, 20 electromechanical unit cells that periodically repeat all along the structure were considered. The unit cell is composed of a portion of the rod with length 50 mm, two entire piezoelectric patches bonded onto the upper and lower surfaces of the rod, centered along the length and poled in opposite through-thickness directions, one (independent) shunt circuit that connects directly the electrodes of these patches and two circuits that connects the pair of patches to left and right neighboring unit cells circuits.

### Methodology and Results

The effect of parametric uncertainties in the vibration mitigation performance of independent resonant shunt circuits and interconnected circuits, focusing on the first vibration mode, was evaluated and compared. First, the nominal case was defined using the optimal values for the circuit components. Then, stochastic models were constructed for the uncertain parameters, considering a Gamma p.d.f., and used for Monte Carlo simulations. Finally, the results are used to assess the robustness of the two solutions.

In this extended abstract, results considering uncertain capacitance for each circuit are shown in Figs. 1 and 2. It is noticeable that, for the interconnected circuit, the mean response and the superior of the confidence interval follows the shape of the nominal response with two peaks, while this is not the case for the independent circuit. In terms of the maximum responses, for the independent circuit, the difference between the maximum values of the nominal frequency response and the superior of the 95% confidence interval is 2.2 dB and 5.4 dB for 5% and 10% dispersions, respectively. For the interconnected circuit, these values are 2.5 dB and 4.5 dB. These results suggest that, for capacitance individual variabilities, the interconnected circuit seems more robust, and thus more interesting, than the independent one.

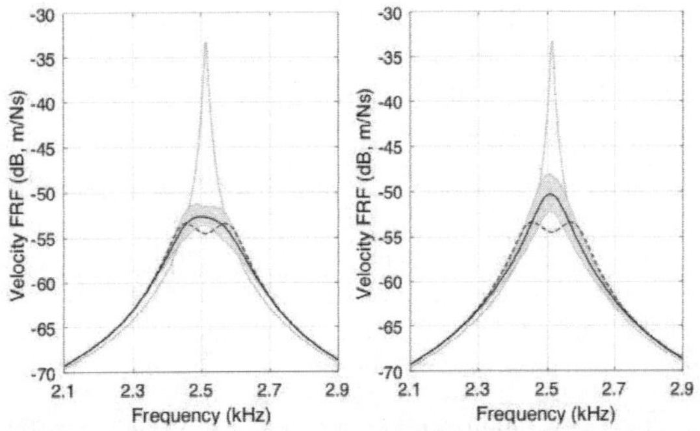


**Figure 1**. Frequency response of rod with independent circuits for uncertain patch capacitance (5% (left) and 10% (right) individual dispersion): open-circuit (short dash), RL nominal (dash), RL mean (solid), 95% confidence interval (fill).

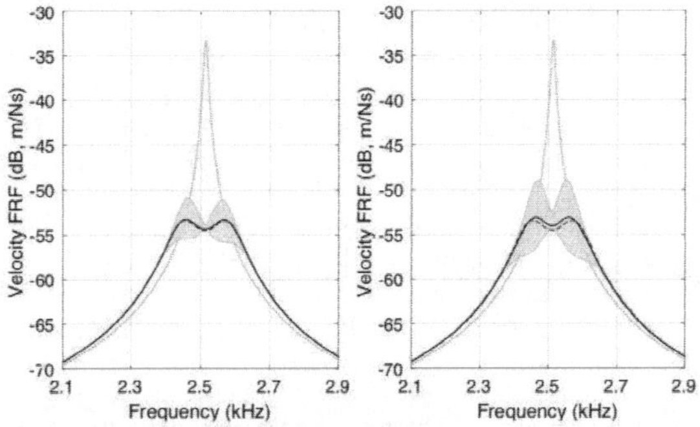


**Figure 2**. Frequency response of rod with interconnected circuits for uncertain patch capacitance (5% (left) and 10% (right) individual dispersion): open-circuit (short dash), RL nominal (dash), RL mean (solid), 95% confidence interval (fill).

Additional analyses considering the effect of uncertain inductance for each circuit and global uncertainties for the capacitances, and also considering the effect of the circuits effective resistances on the robustness of each solution will be also presented.

### References

*Andreaus U and Porfiri M (2007) Effect of electrical uncertainties on resonant piezoelectric shunting. J. Intell. Mater. Syst. Struct. 18:477-485.*

*Lossouarn B, Aucejo M and Deü J-F (2015) Multimodal coupling of periodic lattices and application to rod vibration damping with a piezoelectric network. Smart Mater. Struct. 24:045018.*

*Santos HFL and Trindade MA (2011) Structural vibration control using extension and shear active-passive piezoelectric networks including sensitivity to electrical uncertainties. J. Braz. Soc. Mech. Sci. Eng. 33:287-301.*

# Nonlinear Frequency Analysis of Hair Cell Bundle Structure with Negative Stiffness Characteristic

**Gi-Woo Kim* and Mai Van Ngoc**

*Department of Mechanical Engineering, Inha University, Incheon, South Korea*

**gwkim@inha.ac.kr*

**Abstract**

This study presents the mechanism of nonlinear stiffness and amplification of the sound signal associated with the hair bundle structure. The 3D structure-based bio-inspired hair cell bundle structure which has the nonlinear spring property of negative stiffness is used in this study. Then, the hair cell-based structure is used as nonlinear spring element of 1-DOF vibrating system. Harmonic balance method (HBM), a typical analytical tool used to estimate the nonlinear frequency response functions (FRFs), is applied to analyze the 1-DOF nonlinear mass-spring-damper system. The function of the nonlinearity of human's hair cells is investigated from the HBM analysis results.

**Introduction**

The hair cells inside the human auditory system features a conical bundle structure consisting of numerous tiny long cylinders, which are called stereocilia (Hudspeth, 2005) as shown in Figure 1(a). It has been known that the sensitivity of our hearing can be more increased by the mechano-electrical phenomenon observed inside the hair bundle structures that can mechanically amplify the movements of the hair cells at the low frequency ranges. However, the detail mechanism by which mechanical energy can be amplified remains uncertain. The negative stiffness is consistently observed in the measurement of stiffness of hair bundle structures, and is also known to be harnessed to amplify mechanical auditory stimuli in the hair bundle structures. This study is therefore to explore the mechanism of nonlinear stiffness and amplification of the sound signal. A bio-inspired hair cell bundle structure which has nonlinear spring property such as negative stiffness is used. This nonlinear spring characteristic is then used for the spring element of 1-DOF vibrating system. Harmonic balance method (HBM) will be applied to analyze the 1-DOF nonlinear mass-spring-damper system.

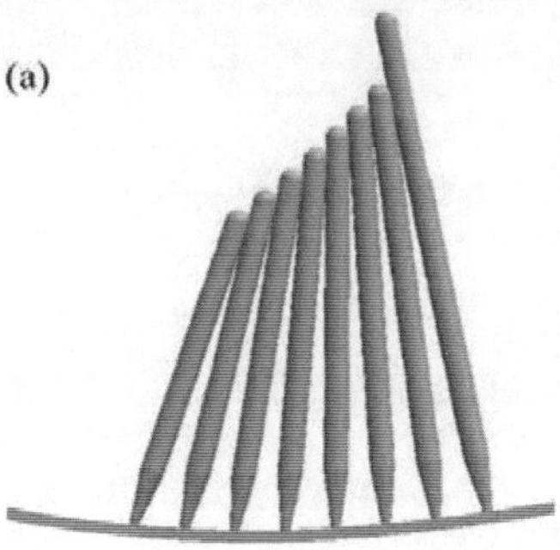


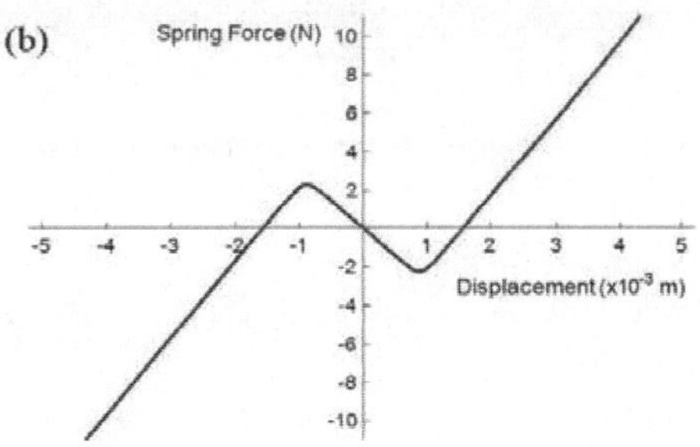


**Figure 1**. A schematic of hair cell structure: (a) a bio-inspired hair cell structure (b) mechanical property

**Methods/Model Formulations**

As shown in Figure 1(b), the nonlinear stiffness function of the bio-inspired hair cell bundle is a curve including 2 identical positive and one negative stiffness coefficient. The nonlinear dynamic behavior of 1-DOF vibrating system employing biomimetic hair cell is investigated by using the HBM method (Yoon and Yoon, 2014). The equation of motion of the vibrating system can be formulated as follows

$$m\ddot{x}(t) + c\dot{x}(t) + F_s(x) = F_e(t) \tag{1}$$

where $m$ is the proof mass (0.015 kg), $c$ is the damping coefficient, $F_s(x)$ indicates the nonlinear spring force induced by the hair cell bundle and $F_e(t)$ is the sinusoidal excitation force. The solution associated with super-harmonic terms caused by the nonlinearity is expressed as follows

$$x(t) = x_m + \sum_{k=1}^{N_{max}} \left[ x_{a(k)} \cos(k\omega t) + x_{b(k)} \sin(k\omega t) \right] \tag{2}$$

Where $k$ are the super-harmonic indices, and $N_{max}$ denote the maximum number of harmonics

**Results and Discussions**

Figure 2 shows that the RMS displacement of the proof mass significantly increases in low frequency region while its amplitude largely decreases at high frequencies compared to a system with a linear spring force when the damping of the system increases. It means that sensitivity of the vibrating system can be more increased by using the negative stiffness of the hair bundle structures that can mechanically amplify the movements of the hair cells in the low frequency region. It is interesting that the 1-DOF vibrating system associated with hair cell structure can be used as a vibration isolator because it has low vibration displacement at high frequencies as shown in Figure 2.

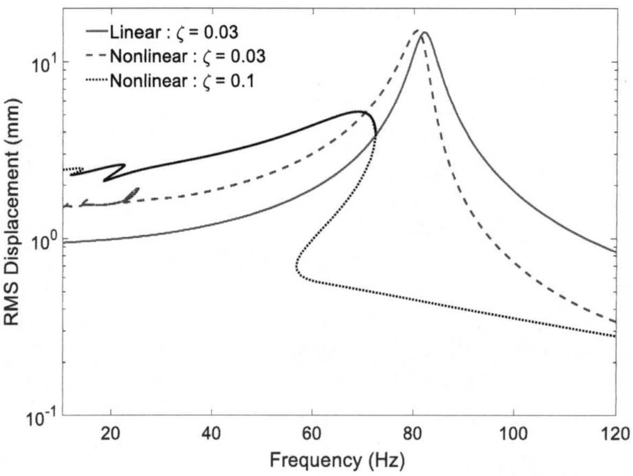


**Figure 2**. Simulation of frequency responses of mechanical vibrating system with negative stiffness

**References**


*Hudspeth AJ (2005) How the ear's works work: mechanoelectrical transduction and amplification by hair cells. Comptes rendus biologies, 328(2): 155-162.*

*Yoon JY and Yoon HS (2014) Nonlinear frequency response analysis of a multistage clutch damper with multiple nonlinearities. Journal of computational and nonlinear dynamics, 9(3), 031007.*

# Shock Reduction on a Thin Plate using Elastic Wave Dispersion Patch

**Dae-Hyun Hwang and Jae-Hung Han***

*Department of Aerospace Engineering, KAIST, Daejeon, 34141, Republic of Korea*

**jaehunghan@kaist.ac.kr*

**Abstract**

Space structures are designed and built using frames and thin plate structures to reduce the overall weight. High-frequency stress wave, which is typically induced by pyrotechnic operations during space missions, propagates through thin plates and frames, and this can cause fatal malfunction in onboard electronic components. In this study, we propose a simple method to disperse a transient bending wave on a thin plate structure using elastic patches.

**Introduction**

The velocity of wave in a thin plate structure is determined by the material properties and the thickness of the plate. In the wave theory, the wave propagating from one medium to the other is refracted by the propagation velocity difference, and this refraction phenomenon is similarly applied to the shock wave on the plate (Kolsky, 2012). Darabi et al. designed and experimentally tested a wave-bending structure in a thin plate by smoothly varying the rigidity of the plate (Darabi, 2018). The corner-shaped trajectory channel was machined on a plate, so that an incident bending wave follows the trajectory. In this study, we changed the velocity of a bending wave propagating on a flat plate by attaching a thin elastic patch on the thin plate structure, and we investigated how to reduce the shock transmission for a specific region by using the dispersion phenomenon caused by the refraction of the bending wave. A numerical analysis using explicit FEM method was used to visualize the process of refraction due to the elastic patch, and the amount of shock reduction with and without the elastic patch was quantified.

**Analysis Model**

ANSYS AUTODYN was used to simulate the propagation of the shock wave in a thin plate. The plate is made out of aluminum (AL-6061) with 5 mm thickness, and the elastic patch has a right-angled triangle shape, and made of the same material. The numbers of total element and nodes are 277540 and 350550, respectively. To enhance the calculation efficiency, the half of the geometry was modeled using the symmetry, as shown in Fig. 1. The shock force input was set as sine wave of 10 kHz.

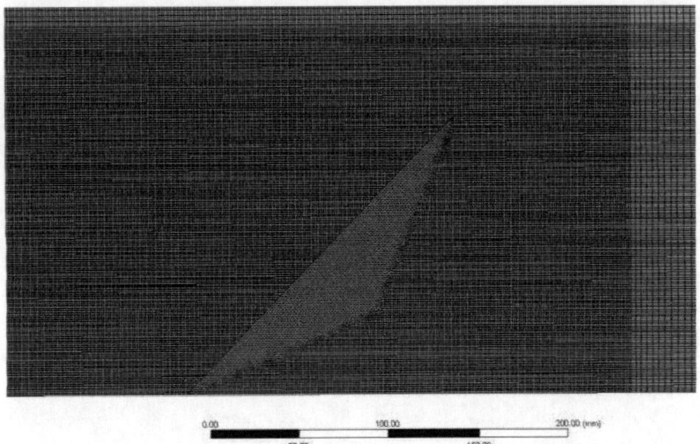


**Figure 1**. Geometry and finite element model of the plate and the elastic patch

## Results and Discussions

Fig. 2 shows the refraction behavior of the incident shock wave. The shock transmitted to the central line behind the elastic patch was significantly reduced; the velocity response was reduced by about 40% compared to the wave passing through the side of the patch.

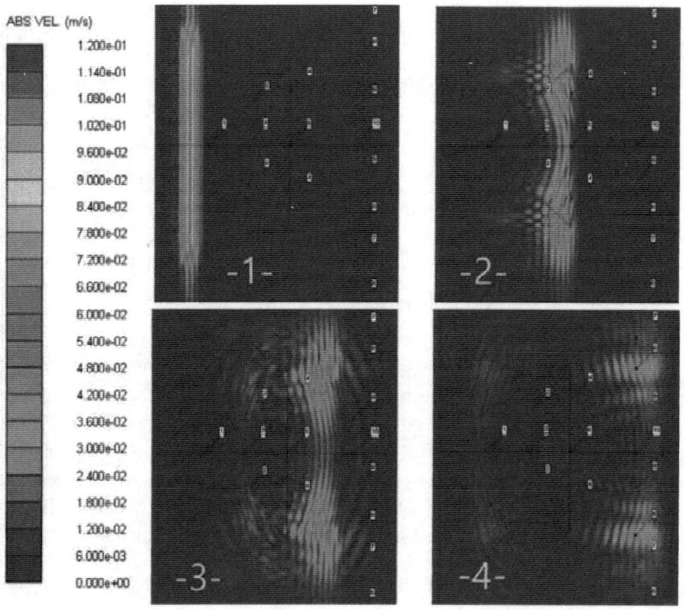


**Figure 2**. The simulation results showing propagated and reflected bending wave through elastic patch

## References

*Kolsky H (2012) Stress Waves in Solids. 2nd ed. Dover Publications, New York.*

*Darabi A, Zareei A, Alam M-R, and Leamy MJ (2018) Broadband Bending of Flexural Waves: Acoustic Shapes and Patterns. Scientific Reports 8:11219*

# Digital Twin Model Proposal for an Automotive Chassis

**Jorge de Jesús Lozoya-Santos[1*], Andrés Campos Ferreira[1], Adriana Vargas Martínez[1], Juan Carlos Tudón-Martinez[2], Ruben Morales-Menendez[1], and Ricardo A. Ramirez-Mendoza[1]**

[1]*School of Engineering and Sciences, Tecnologico de Monterrey, Mexico*

[2]*School of Engineering and Technology, Universidad de Monterrey, Mexico*

[*]*jorge.lozoya@tec.mx*

**Abstract**

The digital twin is a comprehensive representation of a product or process into a digital world, playing a vital role in the product lifecycle and helping to the horizontal integration of the production process. Nevertheless, digital twin technology has not been completely implemented on vehicle systems. This is an opportunity to propose a service digital twin model based on a smart product reference. In this work, a first validation of the architecture, simulation, and experimentation will be studied on a vehicle chassis case of study. Calculations for vehicle dynamics, torsion and strain phenomena will be part of the study case. A smartphone is used to acquire the vehicle dynamics as the most suitable and inexpensive sensing technology.

Also, simulation using commercial software and result validation under certain conditions will be provided.

**Introduction**

According to David Cearley (2018), in the context of industry 4.0 Gartner considers the Digital twin, (DT) as one of the top ten technology strategic trends. Haag and Anderl (2018) define DT as a comprehensive representation of a product or process into a digital world. This allows the DT to perform active participation from the design to the functional stage on each product or process. This work will focus on smart products more than smart processes; however, the DT can be applied to both systems. Nowadays, there are only a few available scientific documents of implementation DT on vehicles. This is an opportunity to move towards industry 4.0 by improving the product with edge technologies. The problem to solve is to obtain accurate and inexpensively the reference data from the system and replicate its behavior on a virtual model.

**Digital Twin Proposal and Methodology**

To generate a DT is necessary to collect a great amount of data from the physical system or product in a cost-effective way. To prove the feasibility of DT with costless data sensing, a service DT focused on a vehicle chassis with a smartphone as a data-collecting device is proposed. The smartphone is a suitable noninvasive device which can measure data as a part of the system, for instance, the vehicle accelerations, turns, velocity, position. Also, due to its versatility and connectivity, an app can be used to retrieve and send to the cloud all the sensor data automatically, Fig. 1a. To prove this proposal, the first simulation connectivity will be done. Retrieving vehicle dynamics from a smartphone and show the dynamics impact on the modeled subsystem. Based on the DT model proposed, a chassis of a vehicle was chosen. This paper focuses on validating data retrieving and connection with a CAD/CAE commercial software. The vehicle through movement will move the smartphone and the sensors will capture the data and sent to a computer where the information will be displayed on a digital model of the chassis. The methodology is as follows, a first simulation was made of common driving events on CarSim software then, the resulting dynamics were used to make a simulation on a CAD/CAE software, Fig. 1b. Later, two experiments were guided by sensing the vehicle dynamics with one and two smartphones on a bumper event, also the results were used to make a CAD simulation, Fig. 1c.

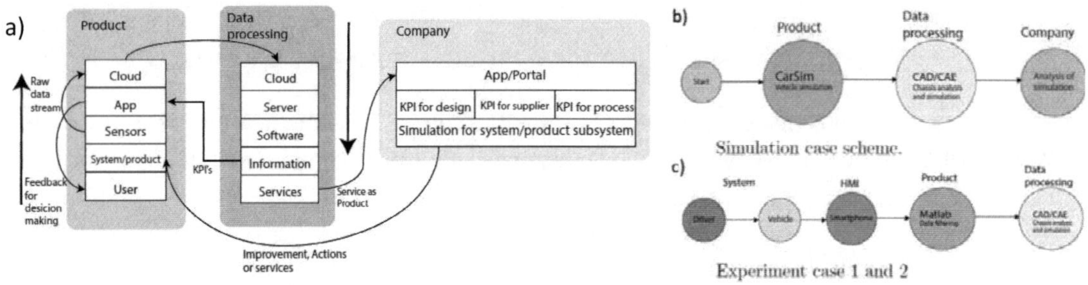


**Figure 1**. Digital Twin: a) communication model, b) simulation scheme, c) experimental scheme.

## Results and Discussions

The vehicle simulation was be done by CarSim (version 8) to generate dynamics data. The accelerations on x, y and z directions, and the rotational speeds retrieved from the simulation are shown in Fig. 2 (left and center). With SolidWorks (version 2017) software, the alteration in the dynamics are notorious, Fig.2 (right), because a bumper was presented. The axis with the main change was the z-axis and also a twirl around y-axis (pitch) rotation.

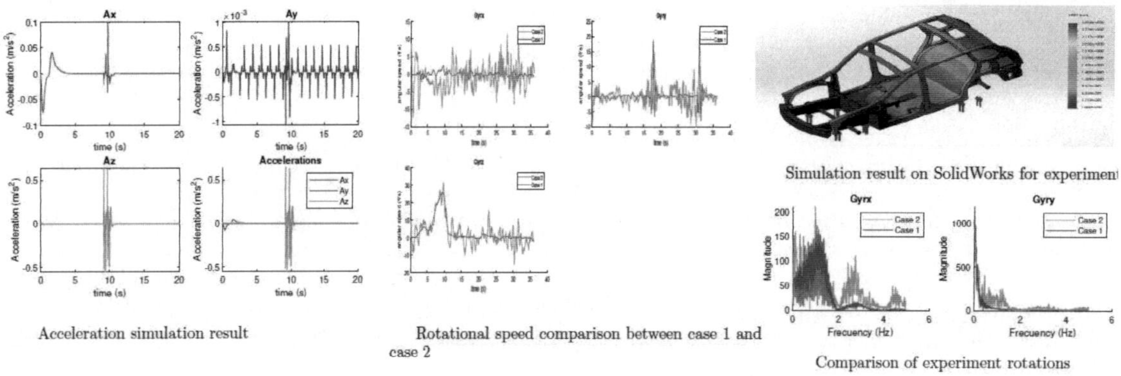


**Figure 2**. Evidence of first results. The first four plots show the simulation results in Carsim. The second group shows the results from experiments in an actual vehicle. The last figures show the obtained simulation with SolidWorks using the registered data. Last figures evidence how the frequency contents in simulation for the pitch, roll, and yaw in the vehicle can achieve good results in structure effect simulation.

The analysis shows the deformation caused by the bumper simulation on Carsim. The deformation is caused by the vertical forces applied on each wheel provoking a deformation mainly on the car body center. This first approach for a DT helps to understand the behavior and aspects to consider during its development.

## References

*David Cearley, Alfonso Velosa, M.K.M.W.B.B. (2018). Top 10 strategic technology trends for 2018: Digital twins. URL{https://www.gartner.com/}.*

*Haag, S. and Anderl, R. (2018). Digital twin proof of concept. Manufacturing Letters, 15, 64{66.*

**Gap in pagination due to unavailable papers.**

**Pages 31-32**

# Multi-mode Piezoelectric Shunt Damping of Plate-like Structures

**Johan Frederik Toftekær* and Jan Høgsberg**

*Department of Mechanical Engineering, Technical University of Denmark, Kgs. Lyngby, Denmark*

*jotof@mek.dtu.dk

**Abstract**

Excessive resonant vibration of flexible plate-like structure may be mitigated by the attachment of shunted piezoelectric absorbers. The resonant shunt circuit consists of an inductor and a resistor tuned with respect to a target resonant frequency of the structure. When several shunted piezoelectric absorbers are attached to the vibrating structure more resonant vibrations may be mitigated simultaneously. A calibration procedure for the tuning of several shunted piezoelectric absorbers is demonstrated experimentally for a free beam and a plate structure with five piezoelectric transducers simultaneously mitigating four resonant vibration modes. Experimental results correlate well with numerical results.

**Introduction**

Flexible plate-like structures may be vulnerable to excessive vibrations caused by dynamic harmonic loads with frequencies close to a resonance in the structure. This may cause structural failure, fatigue or lead to undesirable acoustic problems. The introduction of resonant absorber devices calibrated to a single or several dominating resonant modes of the structure may therefore be required. Regarding vibration mitigation of plate-like structures, a convenient absorber device consists of co-located piezoceramic patches with strong electromechanical coupling properties wired to a resonant shunt circuit, (Toftekær et al., 2019). The resonant shunt circuit consists of an inductance ($L$) tuned to secure a proper absorber frequency and a resistor ($R$) chosen to obtain optimum energy dissipation around the targeted resonant frequency.

**Method and Model Formulation**

The calibration of $LR$ shunts is typically based on the assumption of a modal reduced system, representing the full structural dynamic response around the targeted resonant frequency. This assumption leads to inaccuracies in the corresponding shunt tuning when spill-over from non-resonant vibration modes is neglected. A recent piezoelectric shunt tuning procedure (Toftekær and Høgsberg, 2019) is based on the absorber response given as, respectively, the measured electric current and voltage associated with the short- (SC) and open circuit (OC) piezoelectric interface electrodes, inherently containing the contribution from non-resonant vibration modes, see Tab. 1. This method is a refinement of the method proposed in Toftekær et al., (2019) and is verified experimentally. Regarding the numerical implementation, two eigenvalue problems associated with SC and OC piezoelectric interface electrodes are evaluated, while the experimental implementation requires two frequency response functions (FRF) to be determined from the electric current and voltage response to random dynamic loading of the tested structure. The numerical method is presented for single- and multi-mode tuning in Toftekær and Høgsberg (2019).

**Table 1.** Resonant parallel shunt tuning of $n_p$ circuits with respect to target mode $r$ of piezoelectric transducer $i$ and the corresponding squared electromechanical coupling coefficient $\kappa_r^2$, damping ratio $\zeta_r$ and modal mass $\tilde{m}_r$

| | | | |
|---|---|---|---|
| $L_i^p = \dfrac{\lvert\hat{V}_r\rvert_i}{\lvert Q_r\rvert_i \omega_r^2}$ | $R_i^p = \dfrac{\lvert\hat{V}_r\rvert_i}{\lvert Q_r\rvert_i \omega_r}\sqrt{\dfrac{1}{2\kappa_r^2}}$ | $\kappa_r^2 = \sum_{i=1}^{n_p} \dfrac{\lvert Q_r\rvert_i \lvert\hat{V}_r\rvert_i}{\tilde{m}_r \omega_r^2}$ | $\zeta_r = \sqrt{\dfrac{1}{8}\kappa_r^2}$ |

The current experimental setups, concerning a free aluminum beam Fig. 1(b) and plate Fig. 1(c) with five pairs of collocated piezoceramic patches verifies the optimum multi-mode shunt tuning procedure in Toftekær and Høgsberg (2019). In the experimental setup in Fig. 1(a), the power amplifier excites the structure by a single pair of piezoceramic patches, while the Laser Doppler vibrometer measures the mechanical velocities, from which dynamic amplifications, frequencies and mode-shapes may be determined. Finally, the multimeter measures the electric current and voltage response of the four remaining pairs of piezoceramic patches. The optimum piezoelectric shunt tuning in Tab. 1 is then determined from the SC and OC frequencies ($\omega_r$ and $\widehat{\omega}_r$) and modal charge $|Q_r|_i$ and voltage $|\hat{V}_r|_i$ estimated from the FRFs for the electric current and voltage, respectively. The shunt circuits are then implemented by an inductance realized by winding a copper wire around a magnetic coil connected to a variable resistor.

**Figure 1**. Experimental setup (a) and close-up of the tested free (b) beam and (c) plate with five pairs of co-located piezoceramic patches.

## Results and Discussions

The FRFs for charge and voltage of four of the piezoceramic patch pairs attached to the free beam and plate are attained from the time records to random vibration excitations induced by the patch pair connected to the power amplifier. From the FRFs, see Fig.2, the tuning parameters required in Tab. 1 are obtained and the corresponding shunt tuning is determined for all resonant modes in the considered frequency spectrum.

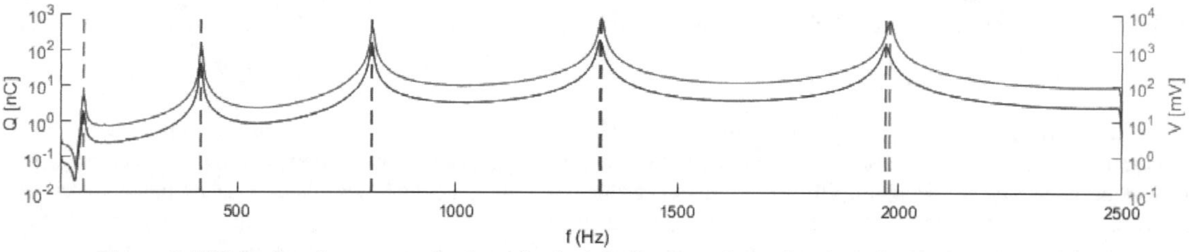


**Figure 2**. FRF for electric current and voltage for the left piezoceramic patch pair on the free beam Fig. 1(b)

The calibration procedure (Toftekær and Høgsberg, 2019) is found to perform well for both single and multi-mode vibration mitigation and the results are in good correlation with the results obtained with corresponding numerical models build in ANSYS® with 3D finite elements (Toftekær and Høgsberg, 2019). While the tuning of the shunt inductance correlates well, the resistance tuning is more uncertain due to lower electromechanical coupling coefficients and parasitic damping in the experiment from mainly a non-ideal gluing of the piezoceramic patches.

## References


*Toftekær JF, Benjeddou A and Høgsberg J (2019). General numerical implementation of a new piezoelectric shunt tuning method based on the effective electromechanical coupling coefficient. Mechanics of Advanced Materials and Structures, doi: 10.1080/15376494.2018.1549297*

*Toftekær JF and Høgsberg J (2019) Independent single- and multi-mode shunt calibration of multiple piezoelectric domains on vibrating structures. (in preparation).*

# Investigating the Performance of Multiple Tuned Mass Dampers on Long Span Cable Stayed Bridge under Multiple Support Excitation

**Zahrasadat Momeni*** **and Ashutosh Bagchi****

*Department of Building, Civil & Environmental Engineering, Concordia University*

*Montréal, QC, H3G 1M8, Canada*

**zahra.momeni24@yahoo.com, **ashutosh.bagchi@concordia.ca*

## Introduction

Cable stayed bridges are characterized by a long fundamental period; and are also flexible and lightweight structures with a low structural damping; a combination which makes this type of bridges vulnerable in large amplitude vibrations due to earthquake or wind excitation[I.E. Haik et all 2005]. Differences in amplitude and phase of seismic motions recorded over long distance is known as spatial variability of ground motion. This phenomenon is proven to have detrimental effect on the seismic response of lifelines such as bridges [Aspasia Zerva, Vassilios Zervas 2002]. On the one hand, Tuned Mass Dampers or vibration absorber is a passive device consisting of a mass, spring and a dashpot attached to a main system calibrated to be resonance with a specific mode of the structure, with the aim of reducing its associated dynamic response [Soong and Dargush 1997]. On the other hand, for controlling the vibration of the systems with variable frequencies, researchers proposed the multiple tuned mass dampers [Zuo and Nayfeh 2005]. In this study the performance of multiple tuned mass dampers for long-span cable stayed bridges under multiple support excitations is investigated. 12 TMDs are assigned to the bridge, in both vertical and transversal direction. The ABAQUS software is used for modeling and dynamic analysis of the considered cable-stayed bridge. The artificial earthquake records is generated taking into account the distance between the piers of the considered bridge. Dynamic analysis is carried out by applying the displacement time histories to the supports of the bridge. In general, the performance of TMDs is satisfactory.

## Methodology

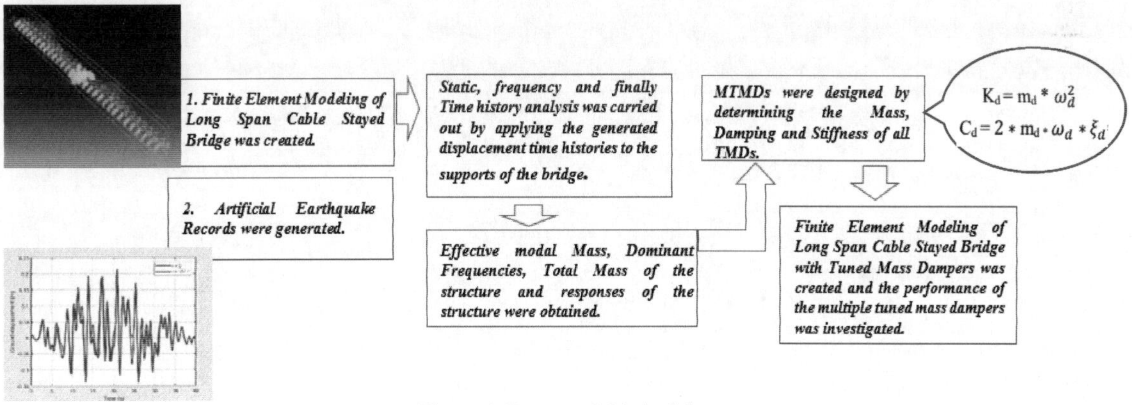


**Figure 1.** Framework Methodology

## Result

The purpose of this study is to investigate the performance of MTMD in long span cable stayed bridge under multiple support excitation. The ABAQUS software is used for modeling and dynamic analysis of the considered cable-stayed bridge with and without multiple tuned mass dampers. The responses of the bridge with and without MTMDs including base shear forces, stress of the cables and the moments of the deck are compared. The base shear forces, the stresses of 43 out 56 cables and moments of the most of the elements of the deck were reduced after installing MTMDs.

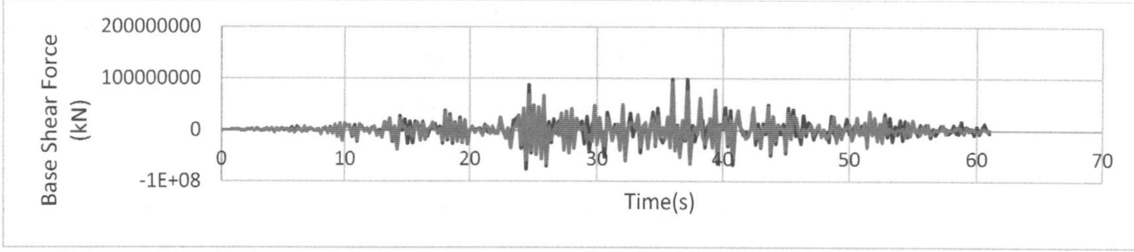


**Figure 2.** Base shear Forces of one of the pier

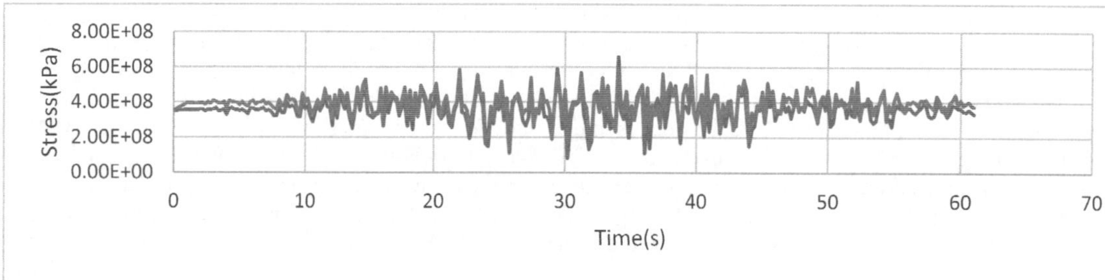


**Figure 3.** Stress of one of the cable

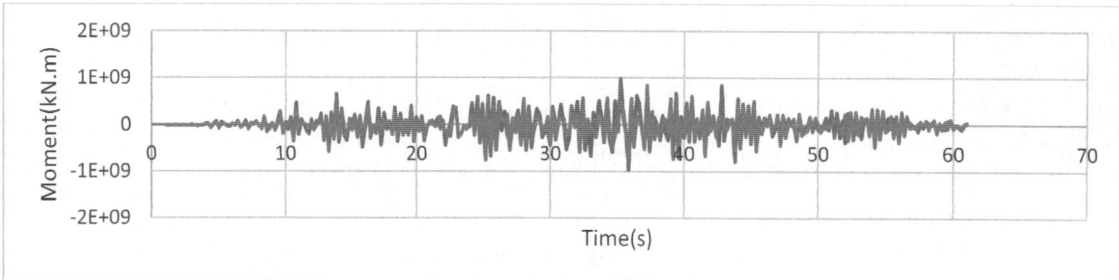


**Figure 4.** Moment of one element of the deck

## References


*I.E. Haik,:"Baseline modeling of the Maysville cable stayed bridge over the Ohio river,"kentucky transportation of engineering college of engineering*

*Zerva A, Zervas V. Spatial variation of seismic ground motions. Applied Mechanics Reviews 2002; 55(3):271–297*

*Zuo L and Nayfeh SA (2005) Optimization of the individual stiffness and damping parameters in multiple-tuned-massdamper systems. ASME Journal of Vibration and Acoustics 127: 77–83*

# Seismic Performance of Concrete Moment Resisting Frames in Western Canada

**Ali Naghshineh*** **and Ashutosh Bagchi****

*Department of Building, Civil and Environmental Engineering Concordia University, Montreal, QC, Canada, H3G 1M8.*

*ali.naghshineh@mail.concordia.ca, **ashutosh.bagchi@concordia.ca

**Abstract**

This paper presented investigation of the seismic performance of a set of code-designed 4, 8 and 12 story moment resisting concrete frames at different hazard levels namely, SLE (Service Level Event), DLE (Design Level Event) and MCE (Maximum Considered Event). Nonlinear static pushover analysis and dynamic time history analysis using a set of ground motion records such as crustal and subduction earthquakes have been used to investigate their effects, the different performance levels of these buildings, and adjust their design based on the corresponding target displacements.

Keywords: seismic performance, performance-based design, concrete moment resisting frame, crustal earthquakes, subduction earthquakes.

**Introduction**

Seismic design codes presently in use are prescribed-based and focus on strength and capacity of structural members, but the structure's overall performance during a given seismic event cannot be clearly described. In this study the seismic performance of a 4, 8 and 12 story moment resisting concrete frames were evaluated based on the FEMA 356, ATC 72 and ASCE 41 procedures.

**Design of Structural models**

The 2015 NBCC was used to design four-, eight-, and twelve-story, which consist of 3-bay by 3-bay ductile moment resisting concrete (DMRF). Detailing followed the Canadian Standard for reinforced concrete buildings (CSA A23.3). The buildings are assumed to be located in Victoria, BC, in the western part of Canada with site class "C". A quasi static member with inelastic concentrated hinges for a reinforced concrete flexural member was used for modeling parameters namely as pre-capping, $\theta_p$ and post capping $\theta_{pc}$ plastic rotations as well as the cyclic deterioration $\lambda$ based on FEMA 356 and ATC 72, and calculated for all the beams and columns. Then a nonlinear static analysis is performed for all the three models, this process involves pushing the structure under monotonically increasing lateral loads until a target displacement is exceeded. An idealized relationship between base shear and displacement was created for all the three models and the elastic lateral stiffness $K_i$, effective lateral stiffness $K_e$ as well as the effective yield strength $V_y$ are obtained, which presented in Figure 1a only for a 4-story. The target displacement for building with rigid diaphragm was calculated based on FEMA-356/ASCE-41 for SLE, DLE, and MSE, the results of pushover curves are shown in Figure1b, 1c, and 1d.

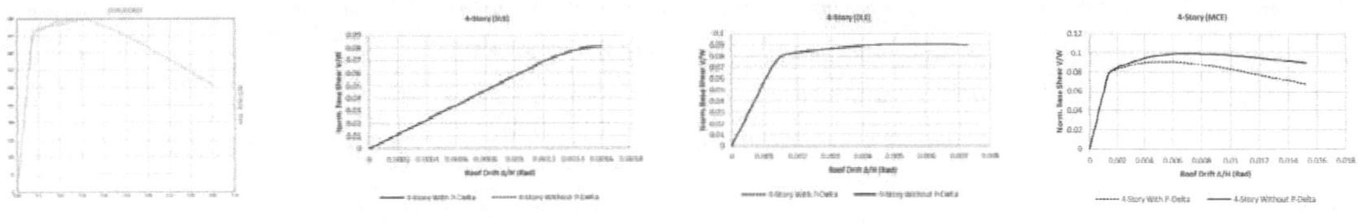

(a) (b) (c) (d)

**Figure 1**. (a) Equivalent bilinear curve, (b), (c), (d) Normalized base shear vs. roof drift ratio for SLE, DLE and MCE for 4-story.

## Roof, inter-story demand results

In this part, 8 different ground motion records are selected and scaled according to the design spectrum. Figures 2 shows the story drift ratio for scaled ground motions for a 4-story as well as SDR for different MCE, DLE and SLE of all the three models. As it was expected, the structure remains elastic in SLE case. It can be observed from the sum of the standard deviation and the mean values that the inter-story drift is in the range of 1.5% to 2.5% which is the "Life safety" performance of the structure defined in NBCC 2005.

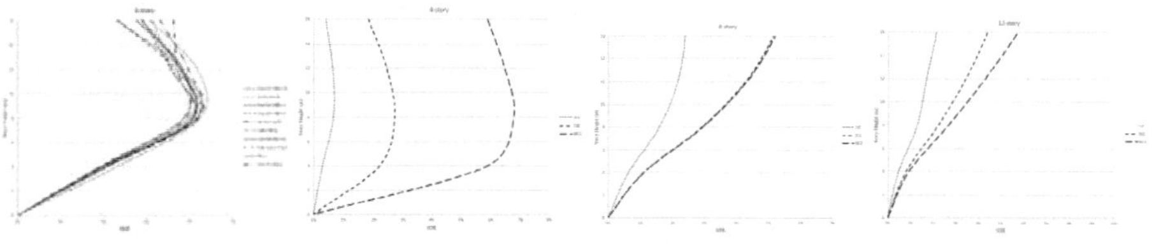

**Figure 2**. Story Drift Ratio for scaled ground motions (4-Story on the left), SLE, DLE and MCE for 4,8,12-Story.

## Results and Discussions

It can be observed, the building members can sustain a damage from earthquake, however if the P-Delta effect is neglected, they will collapse due to gravity load. Also, it was observed the second order effects, geometric nonlinearities, have a significant reduction in the flexural capacity of the columns to consider this phenomenon the strong column-week beam approach shall be performed. Results obtained from push over analysis shows that SLE, DLE remains in elastic part for 4-story but in DLE and MCE due to the second order elastic analysis the structure enters in inelastic part for 8-, and 12-story. It is evident that there is not too much difference between FEMA 356 and ATC 72, however FEMA 356 is more conservative which is based on the skeleton curve, whereas ATC 72 is more reliable due to considering many parameters which is based on monotonic backbone curve. Moreover, for buildings with fundamental period less than 0.8 second, the higher mode effects can be ignored due to the use of load pattern, which gives reliable results for rigid or short structures, therefore for tall buildings a dynamic procedure is required.

## References


*CSA (2014) Design of concrete structures. CSA A23.3-14, Canadian Standards Association, Rexdale, Ontario.*

*NBCC (2015). National Building Code of Canada.*

*Deierlein GG, Reinhorn AM, Willford MR(2010) Nonlinear structural analysis for seismic design, NEHRP seismic design technical brief 4, 1–36.*

*Elwood KJ, Eberhard MO(2006) Effective Stiffness of Reinforced Concrete Columns.*

*Elwood KJ, Matamoros AB, Wallace JW, LehmanDE, Heintz JA, Mitchell AD, Moore MA, Valley, MT, Lowes LN, Comartin CD, Moehle, JP (2007) Update to ASCE/SEI 41 Concrete Provisions. Earthquake Spectra 23, 493–523.*

*FEMA 356 (2000) Prestandard and Commentary for the Seismic Rehabilitation of Buildings*

*PEER Ground Motion Database,PEER Center, https://ngawest2.berkeley.edu.*

*PEER, ATC 72-1 report released about nonlinear modeling of tall building*

*Ahmed Ghobarah (2001) Performance-based design in earthquake engineering: state of development, Journal of Engineering Structures, Vol.23, PP.878–884.*

*ASCE/SEI 41–13. (2014). Seismic evaluation and retrofit of existing buildings, American Society of Civil Engineers, Reston, Virginia, USA.*

# Tailoring the Flexural Response of Sandwich Structures with Faces Made from Shape Memory Alloy Composite Actuators

**Luke Mizzi***, **Andrea Spaggiari, and Eugenio Dragoni**

*Department of Engineering Sciences and Methods, University of Modena and Reggio Emilia, Italy*

*luke.mizzi@unimore.it

## Introduction

Shape memory alloys (SMAs) are materials that 'remember' a preset shape during loading but recover their original conformation upon heating. This property makes them particularly useful for use as actuators, especially in view of the fact that most SMAs are also characterized by a high force/stroke ratio. Typical SMA composite actuators are designed as two component systems incorporating an agonistic-antagonistic relationship such as an SMA strip or wire embedded within an elastic matrix. In this arrangement, the matrix acts as a counterbalance to the SMA component by reversing the actuation stroke of the system upon cooling, thus imparting reusability to the actuator (Spaggiari and Dragoni, 2014). In this work we present an analytical model for designing and optimizing the geometry of an SMA actuator. Based on the geometry and material properties of the individual components, the model allows one to predict the axial stroke and recovery of the actuator. The predictions of this model were validated using Finite Element (FE) simulations (Mizzi, Spaggiari and Dragoni, 2019) where the material properties of the SMA component were described by the numerical Souza-Auricchio model (Souza, Mamiya and Zouain, 1998; Auricchio and Petrini, 2004). Finally, these two approaches were extended to the design of sandwich structures with SMA-matrix faces and soft cores, which can undergo controlled flexural deformations upon selective heating one of the surface SMA components (see Figure 1).

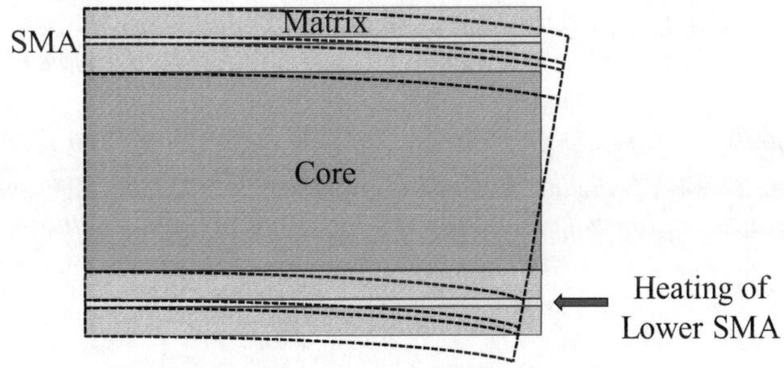


**Figure 1**: Schematic of a bending SMA sandwich structure upon selective heating of one of the SMA components on the surfaces of the system.

## Methods and Model Formulations

The SMA-matrix actuators investigated in this study are assumed to be designed in the following manner (based on an experimental production technique proposed by Bettini et al., (2009); 1) the SMA component is stretched by a fixed level of strain, 2) the matrix is formed around the pre-stretched SMA, which is then 3) released and the system is allowed to equilibrate. The actuator then functions by heating the SMA component to induce an austenitic phase transformation which produces an actuation stroke, followed by re-cooling of the system back to its original temperature which results in the system attempting to return back to its initial state.

Both the analytical model and Finite Element simulations were conducted on the basis of this line of thought. The former was based on the equilibration of forces between the SMA component in its hot and cold states (defined by a linear elastic austenitic plot and a three-stage martensitic plot respectively) and the counterbalancing forces imparted by the matrix (linear elastic). In order to analyze the extent of possible actuation strokes and recovery for these SMA actuators, a range of geometric parameters, pre-strain values and material properties were considered, with particular focus on the stiffness of the matrix and the extent of initial applied pre-strain. These are the two main parameters which may be most easily varied during the design phase. A similar methodology was also employed to investigate the deformation of the sandwich system, however instead of actuation stroke, the deformation was measured in terms of the bending angle, φ.

**Results and Discussions**

The results obtained show that the actuation properties of the SMA composite actuator are heavily influenced by the geometry and material properties of the individual components making up the system. The actuation potential of the system increases upon increasing the amount of pre-strain and decreasing the stiffness of the matrix; however, the actuation stroke may be unrecoverable if the matrix is too soft. Moreover, the analytical and FE results show excellent agreement, highlighting the validity of this approach for optimizing the design of these SMA-based actuators.

These findings were used to design sandwich structures with surface skins made up of these SMA-matrix actuators. It was found that the additional bending stiffness imparted by the sandwich structure reduces the actuation stroke of the SMA component upon heating. However, using the proposed model, it is possible to minimize this effect by adopting an efficient geometric design for the entire system. These results highlight the potential of these systems for use in a number of applications where a repeatable, customized axial-flexural actuation is desired.

**References**


*Auricchio, F. and Petrini, L. (2004) 'A three-dimensional model describing stress-temperature induced solid phase transformations: Solution algorithm and boundary value problems', International Journal for Numerical Methods in Engineering, 61(6), pp. 807–836. doi: 10.1002/nme.1086.*

*Bettini, P. et al. (2009) 'Carbon fiber reinforced smart laminates with embedded SMA actuators-part I: Embedding techniques and interface analysis', Journal of Materials Engineering and Performance, 18(5–6), pp. 664–671. doi: 10.1007/s11665-009-9384-z.*

*Mizzi, L., Spaggiari, A. and Dragoni, E. (2019) 'Design-oriented modelling of composite actuators with embedded shape memory alloy', Composite Structures. Elsevier, 213(January), pp. 37–46. doi: 10.1016/j.compstruct.2019.01.057.*

*Souza, A. C., Mamiya, E. N. and Zouain, N. (1998) 'Three-dimensional model for solids undergoing stress-induced phase transformations', European Journal of Mechanics, A/Solids, 17(5), pp. 789–806. doi: 10.1016/S0997-7538(98)80005-3.*

*Spaggiari, A. and Dragoni, E. (2014) 'Analytical and numerical modeling of shape memory alloy Negator springs for constant-force, long-stroke actuators', Journal of Intelligent Material Systems and Structures, 25(9), pp. 1139–1148. doi: 10.1177/1045389X13493354.*

# 3D Printing to Automate Shape Memory Alloy Placement in Composite Structures

**Sampada Bodkhe[1*], Lorenzo Vigo[2], and Paolo Ermanni[1]**

[1]*Composite Materials and Adaptive Structures Laboratory, Department of Mechanical and Process Engineering, ETH Zurich, Leonhardstrasse 21, 8092 Zurich, Switzerland*

[2]*Materials Science and Engineering Department, EPFL, Station 12, 1015 Lausanne, Switzerland*

**bodkhes@ethz.ch*

**Abstract**

A new 3D printing technique was developed to automate the selective placement of shape memory alloys (SMA) wires in composite structures for shape morphing. The technique that uses a polymer coating to isolate the SMA from the electrically and thermally conducting carbon fibers, works at room temperature in order to avoid any alteration to the programming of the SMAs while printing. Selective placement of SMAs allows for shape morphing with complex shapes and curvatures.

**Introduction**

Shape memory alloys (SMA) are integrated into composite structures for shape morphing, vibration attenuation, and self-healing (Roytburd et al., 2000; Cohades and Michaud, 2018; Sanusi et al., 2014). For shape morphing applications, the SMAs are embedded off the neutral axis to achieve an out-of-plane deformation in the composite structures (Choi and Lee, 1998). Carbon fiber reinforced polymers (CFRP) are a good choice as they have a negative coefficient of thermal expansion that complies with the negative strain of the SMAs. Unfortunately, the electrical conductivity of the carbon fibers leads to Joule heating of the matrix (Cohades and Michaud, 2018). In order to avoid this effect, and increase the service life of the hybrid structures (CFRP + SMA), we coated our SMA fibers with an insulating polymer right before placing them on the CFRP prepregs. 3D printing was used to effectively and simultaneously coat the SMAs and automatically deposit them at the desired locations on the prepregs. The prepreg structures were then cured below the actuation temperatures of the SMAs to avoid any transformation of the SMAs during processing and a post-curing step was added to ensure maximum possible cure of the resin.

**Methods**

A commercial 3-axes robot (TR83400-06, H. Sigrist & Partner AG) was modified to facilitate the printing of shape memory alloy (NiTi) wires (SMA, SmartFlex 02, SAES Getters, diameter = 200 μm) coated with polyetherimide (PEI, Resin 1000, Ultem™) on to unidirectional thin ply prepregs (NTPT, 40 g m$^{-2}$) as seen in Fig. 1a. A solution with 10 wt.% of PEI in chloroform was poured into 3 mL syringe barrels and extruded under pneumatic pressures. For the results mentioned here, two SMA wires were laid 10 mm apart onto 2 plies of carbon fiber prepreg (inset Fig. 1b). The direction of the fibers in the prepreg were perpendicular to the SMA length. A top ply was then placed on the previous assembly and was cured at 70 °C for 10 h in an autoclave under a pressure of 4 bars. The specimen was mounted vertically in the form of a fixed cantilever and digital image correlation (DIC) was used to record the out-of-plane deformations of the CFRP structure with embedded SMAs.

**Results and Discussions**

A new module was designed to print SMAs coated with PEI directly on CFRP prepregs. The PEI concentration and the printing speeds were optimized to obtain a continuous coating. The coating thickness was between 150 – 200 μm, which could be varied by varying the printing parameters. SMAs were placed perpendicular to the fiber direction, to record the maximum possible deflections. Multimeter measurements showed that most of the SMAs were electrically isolated from the matrix. A current of 1.5 A was applied for 3.5 s to actuate the specimen. Fig. 1b shows the deflection data obtained from the DIC: The hybrid structure with 2 SMAs deformed 20% higher than the one with a single SMA (results not shown here).

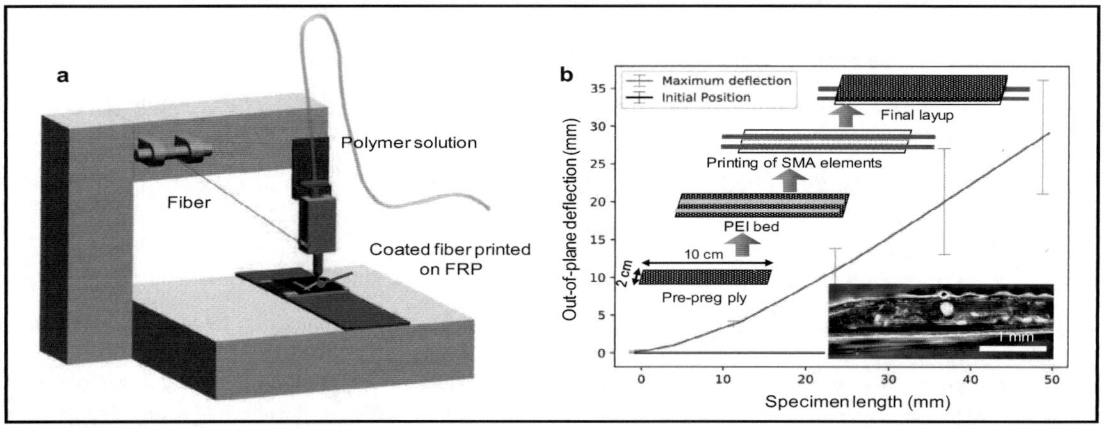


**Figure 1.** a) 3D printer set up for coating and printing the SMA; b) Structure deformation with two embedded SMAs (inset: SMA integration process).

## Conclusions

Although high deflections were attained with the integration of the SMAs with good interfacial bonding (inset Figure 1b) between the fiber and the PEI, and the PEI and epoxy, issues with respect to, complete curing of the composite, experimenting with different layups and the effect of solvent on the FRPs are being investigated. The process developed here can further be employed into 3D printing SMAs in polymer structures for distributed actuation as an improvement over drop-in actuators in biomedical and robotic applications (She et al., 2015; Boyraz et al., 2018)

## Acknowledgements

Nicole Aegerter and Dr. Kress Gerald kindly provided the PEI and the CFRP prepregs, respectively. The authors also thank Prof. Veronique Michaud, EPFL for her valuable suggestions and inputs. The SNF 206021_150729 grant supported the acquisition of the DIC equipment used in our research.

## References

*Boyraz P, Runge G and Raatz A. (2018) An Overview of Novel Actuators for Soft Robotics. Actuators 7: 48.*

*Choi S and Lee JJ. (1998) The shape control of a composite beam with embedded shape memory alloy wire actuators. Smart Materials and Structures 7: 759-770.*

*Cohades A and Michaud V. (2018) Shape memory alloys in fibre-reinforced polymer composites. Advanced Industrial and Engineering Polymer Research 1: 66-81.*

*Roytburd AL, Slutsker J and Wuttig M. (2000) 5.23 - Smart Composites with Shape Memory Alloys. In: Kelly A and Zweben C (eds) Comprehensive Composite Materials. Oxford: Pergamon, 507-522.*

*Sanusi KO, Ayodele OL and Khan MTE. (2014) A concise review of the applications of NiTi shape-memory alloys in composite materials. South African Journal of Science 110: 1-5.*

*She Y, Li C, Cleary J, et al. (2015) Design and Fabrication of a Soft Robotic Hand With Embedded Actuators and Sensors. Journal of Mechanisms and Robotics 7: 021007-021007-021009.*

# Constitutive Model of the Correlation Between the Applied Stress, Strain and Electrical Resistance of SMA Actuator

**Xin Xiang Jiang*, Brian Lynch, Alex Ellery and Fred Nitzsche**

*Department of Mechanical and Aerospace Engineering, Carleton University, Ottawa, ON, Canada*

**xinxiangjiang@cunet.carleton.ca*

## Abstract

The correlation between the electrical resistance and strain of SMA actuator during its phase transformation has been often used as self-sensing method for precision position control of SMA actuator. Existing models primarily focus on the correlation between the electrical resistance and strain of SMA actuators, which have not sufficient taken into account of the influence of applied load stress. Later researches indicate that applied stress significantly influence the correlation between the electrical resistance and strain of the SMA. This paper presents a new approach to the modeling of the resistance-strain correlation of shape memory alloy through introduction of a concept of 3-D modeling of the applied stress, the strain and electrical resistance of the shape memory alloy.

## Introduction

Precision position control of SMA actuators has been a challenging task that continues to be the focus of much research[1,3]. Many early actuators were open-loop systems that most often used on-off control. This type of control is normally sufficient for mechanisms that operate between two states, such as a valve with only open and closed positions. However, most applications require a variable position control of high precision for the actuators in unpredictable variable ambient conditions, which practically can only achieved through a closed-loop control method. While using a position sensor to provide feedback signal for the control is one of the solutions, it brings significant penalties to the actuator design in terms of mass, volume and design simplicity, which defeats main advantage of using SMA actuator. Fortunately, SMA, such as Ni-Ti SMA, exhibits a strong and repeatable correlation between actuator strain and its internal electrical resistance[3], which has been exploited to provide a self-sensing as feedback for position control. There have been some modeling works on correlation between the strain and change in electrical resistance of the SMA actuator [2]. A key factor, which has not been sufficiently taken in to account in those models, is the applied stress, which has been found to influence such correlation significantly. This led to recent establishment of 3-D diagram between strain, change in electrical resist, and applied stress based on experimental data in order to achieve significantly improved control result [3]. However, constitutive equations or theory for such 3-D diagram has not been fully established, which is considered to be needed to understand the theory behind such model and convert it into a practical engineering model for engineers to design the SMA actuators for precision position control purpose. This prompts the current study.

## Methods/Model Formulations

The operating principle of SMA actuator is fundamentally based on the reversible phase transformation between the martensite and austenite phases of SMA material, which have distinctly different electrical resistivities. To establish a constitutive equation governing the variation in electrical resistance of a linear SMA actuator and its strain and applied stress, the first step is to define the parameters of the actuator when it is in full austenite state as the initial parameters (the initial length ($L_0$), strain ($S_0 = 0$), cross-section area ($A_0$) and electrical resistivity of Austenite ($\rho_A$)). When the SMA actuator is being relaxed from the fully actuated state, the phase transformation from Austenite to Martensite will produce a strain (S) and Martensite fraction ($\xi$) in the SMA material. A set of equations governing the relationship among the electrical resistance, strain and stress of SMA actuator can be established based on fundamental physical principles and the relationship between strain (S) and Martensite fraction ($\xi$) in the SMA material:

$$R = \frac{\rho L_0 \, (1+S)^2}{A_0} \tag{1}$$

$$\rho = \rho_A + \xi(\rho_M - \rho_A) \tag{2}$$

$$S = \frac{T}{Y} + \xi \, S_L \tag{3}$$

Here, $\rho_M$ is the electrical resistivity of Martensite of SMA, $S_L$ is the maximum residual strain of the actuator, T is the applied stress. The Y is elastic modulus of the SMA material, which is a variable of $\xi$ in principle, but is treated as a constant for reason of simplification here.

Combining the above three equations, the correlation among the R, S and T can be obtained:

$$R = \frac{L_0}{A_0}[\rho_A + \alpha S^3 + (\rho_A + 2\alpha - \alpha\frac{T}{Y})S^2 + (2\rho_A + \alpha - \alpha\frac{T}{Y})S - \alpha\frac{T}{Y}] \tag{4}$$

Here $\alpha = \frac{(\rho_M - \rho_A)}{S_L}$, which is a constant.

**Results and Discussions**

The constitutive equation (4) demonstrates clearly that the electrical resistance (R) of SMA actuator is a function of both strain (S) and applied stress (T) during its actuation/relaxation. It can be seen that when the applied stress is treated as a constant, the electrical resistance (R) would be a 3rd order polynomial function of strain (S). It may be simplified as quadratic or even linear function depending the requirement of control precision. Subsequently in the case the strain (s) is a constant, the electrical resistance (R) would be simply a linear function of the stress (T). The equation (4) essentially can be used to build a theoretic 3-D model relating the R, S and T for use in high precision control of SMA linear actuator. Validation of this model using experimental data was conducted. The results indicate there is a generally good fitting between the theoretic model and experimental data.

**References**

*[1] Peng, P., X. Jiang, Y. Hu & A. Ng, "Actuation precision control of SMA actuators used for shape control of inflatable SAR antenna", Acta Astronautica, Vol.63, No.5-6, pp578-585, Sept. 2008.*

*[2] Cui D, G Song and H Li "Modeling of the electrical resistance of shape memory alloy wires", Smart Materials and Structures, Vol. 19, No. 5, 2010*

*[3] Lynch, B., X. Jiang, A. Ellery & F. Nitzsche, "Characterization, modeling, and control of Ni-Ti shape memory alloy based on electrical resistance feedback" Journal of Intelligent Material Systems and Structures vol.27 no.18 pp2489-2507, 2016*

# Shape Memory Polymer Composites and 4D Printing Technologies: From the Aerospace to Flexible Electronics

**Jinsong Leng[1*] and Yanju Liu[2]**

[1]*Centre for Composite Materials and Structures, Harbin Institute of Technology, Harbin, China*

[2]*Department of Aerospace Science and Mechanics, Harbin Institute of Technology, Harbin, China*

[*]*lengjs@hit.edu.cn*

**Abstract**

Shape memory polymers (SMPs) and their composites (SMPCs) are prominent stimuli-responsive smart materials that can automatically deform under external stimuli according to the pre-designed setting. The unique advantages of SMPs endow more functions to the devices made of them thus enable novel applications of the products. The speech focuses on the applications of SMPs and 4D technology in some areas, especially applications in flexible electronics and biomedical. Some examples like shape memory bone stent, vascular stent and shape memory polymer-based flexible electronic devices (SMPFEDs) are introduced. Finally, the technical challenges and development directions are proposed.

**Introduction**

In the past few decades, shape memory polymers and their composites have rapidly developed, drawing tremendous attention from various fields. Different from the traditional materials, SMPs/SMPCs have the benefits of fast response, high resilience, light weight, low cost, variable modulus and easy processing, making the the most promising materials for various smart structures and devices and the emerging materials for flexible electronics.

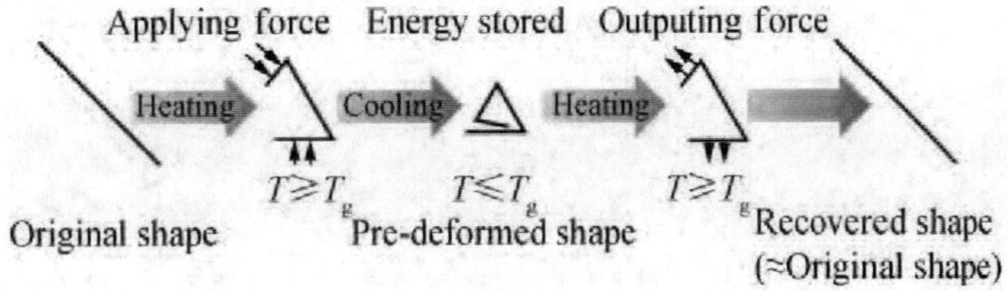


**Figure 1**. Shape recovery mechanism of the thermal-responsive shape memory polymer (SMP).

4D printing refers to a novel additive manufacturing that adds another dimension—the time to 3D printing, which makes the shape or volume transformation time-dependent. The 4D technology enables rapid development of "living" reconfigurable structures at low cost, making it possible to fuel the development of smart products and lead to the growth of opportunities in aerospace, medical and other specialties.

Our group is focused on advanced 4D printing in SMPs/SMPCs. A series of 4D printed SMPs/SMPCs structures with shape or size deformations in response to external stimulus have been developed, such as bone stent, vascular stent and space deployable structures. Through selective incorporation of functional fillers in defined regions, remote actuation of these 4D printed structures with light, electricity and magnetic field have been realized. Multistage, reversible and programmable structures have also been prepared via 4D printing, and systematically investigated and discussed. Going forward, 4D technology will continue to grow and cause revolutionary in many crucial areas. In the final, challenges and opportunities for future research are addressed.

## Results and Discussions

The shape memory effect and other unique features of SMPs leads to new research on SMP-based devices or products in traditional areas. Bone or vascular stent and some optical devices have been developed based on different SMP substrates. The novel 4D printing technology is promising to build more complicated structures and functions, contributing to a wide range of application of SMPs. However, there're still a number of challenges in commercial production or practical applications of SMPs. With broad transition temperature ranges, large modulus spans and two-way or multiple shape memory effect, SMPs are obviously potential materials for both traditional areas and emerging areas.

## References


*Gao, H., Li, J., Zhang, F., Liu, Y., & Leng, J. (2019). The research status and challenges of shape memory polymer-based flexible electronics. Materials horizons,.*

*Gao, H., Huang, L., Liu, Y., & Leng, J. (2018). Developments and challenges of shape memory polymers in flexible optoelectronics. Acta Materiae Compositae Sinica,35(12): 3235-3246.*

*H. Wei, Q. Zhang W, Y. Yao, et al. (2017).ACS Applied Materials & Interfaces, 9: 876-883, 2017.*

*Leng.J, Lan.X, Liu.Y, et al.(2011) .Progress in Materials Science, 56: 1077-1135.*

*S. Miao, N. Castro, M. Nowicki, et al.(2017).Materials Today, 20: 577-591.*

# Impact of Thermally Sensitive Self-Adaptive Structures on the Performance of a Cooling Device

**M. Vilarrubí[1], J. Rosell[1], L. G. Fréchette[2], and J. Barrau[1*]**

[1]*Dynamic Systems Applied to Solar Energy Research Group, University of Lleida, Lleida, Spain*

[2]*UMI-LN2, Institut Interdisciplinaire d'Innovation Technologique (3IT), Université de Sherbrooke, Sherbrooke, QC, Canada*

[*]*jerome.barrau@udl.cat*

**Abstract**

This study demonstrates experimentally the impact of thermally sensitive self-adaptive structures of the performance of a cooling device submitted to non-uniform and time dependent heat load scenarios.

**Introduction**

Cooling devices have been designed during the last decades in order to meet, basically, the requirements of high compacity and low thermal resistance (Sharma, 2015). Additional capabilities, such as low pumping power and temperature uniformity have been also object of study (Riera, 2015) due to their impact on both the efficiency and the reliability of the cooled devices. Researchers have developed procedure to optimize the design of the cooling devices, with respect to different criteria (or combination of them), for a given heat load (uniform or not). However, most of the applications present transient heat load scenarios. In order to optimize the cooling performance along a time dependent heat load scenario, the cooling device should integrate a self-adaptive internal geometry.

This study presents the experimental validation and the impact assessment of thermally sensitive self-adaptive fins integrated in a cooling device, with the aim of improving its temperature uniformity when submitted to non-uniform and time dependent heat flux scenarios.

**Methods**

Shape Memory Alloys (SMA) wings were trained to obtain the Two-Way Shape Memory Effect (TWSME). In this study, the SMA is Nitinol, with the dimensions and properties described in Table 1.

**Table 1**. Austenite and Martensite phase transformation properties and dimensions of the Nitinol sample

| Dimensions (mm) | | | Temperatures (°C) | | | | |
|---|---|---|---|---|---|---|---|
| **Thickness** | **Width** | **Length** | **As** | **Ms** | **Af** | **Mf** | $\Delta T_{Af\text{-}Mf}$ |
| 1.2 | 10 | 30 | 95 | 90 | 112 | 70 | 42 |

The coolant (water) flows through a channel where the trained wing has been integrated (Fig. 1).

At high temperature (Austenite phase), the fin is raised in order to generate vortex and, thereby, a heat transfer enhancement, while it remains in a flat shape at low temperature (Martensite phase) in order to reduce the pressure losses when the heat extraction requirements are low.

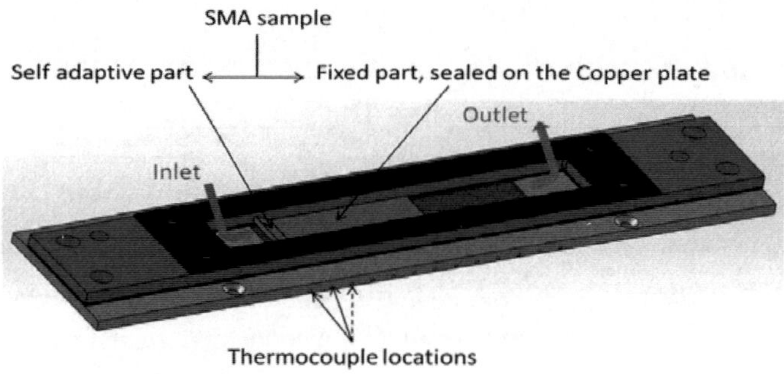


**Figure 1**. 3D view with the SMA wing raised.

## Results and Discussions

The reduction in thermal resistance of the cooling device, obtained through the use of the TWSME of SMA wings, allows maintaining a surface temperature nearly constant even though the heat flux increases (Fig. 2). For fixed inlet coolant temperature and flow rate, the cooling device is able to maintain the temperature within an interval of 7°C for heat flux variations from 32.8 to 67 W/cm$^2$, compared to an expected increase of more than 85°C without self-adaptation.

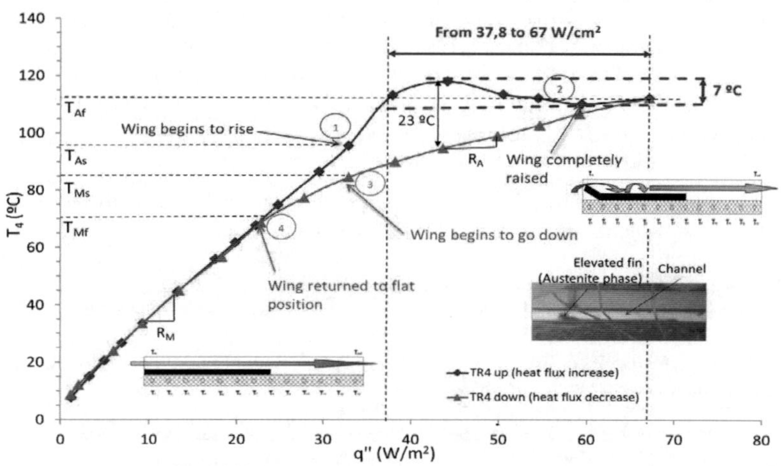


**Figure 2**. Temperature evolution under varying heat flux, for a variation from 0 to 67 W/m$^2$ and back down to 0 (Sample 2, Q=320 ml/min and $T_{inlet}$ = 3°C).

## References

*Riera et al. (2015) Stepwise varying width microchannel cooling device for uniform wall temperature: Experimental and numerical study. Applied Thermal Engineering 78: 30–38.*

*Sharma CS et al. (2015) Energy efficient hotspot-targeted embedded liquid cooling of electronics, International Journal of Heat and Mass Transfer 88: 684–694.*

# Training for Adaptive Three-Dimensional SMA Bumps with Two-way Shape Memory Effect

**Chen Zhang, Hongli Ji, Lin Hao, and Jinhao Qiu***

*State Key Laboratory of Mechanics and Control of Mechanical Structures, Nanjing University of Aeronautics and Astronautics, Nanjing 210016, China*

**qiu@nuaa.edu.cn*

**Abstract**

Two typical and effective approaches for training a 3-D shape memory alloy (SMA) bump with clamped boundaries are studied in present work. Unique training set-up which can be connected to the universal testing machine is designed to obtain the material's two-way shape memory effect (TWSME). After training, SMA bumps with the TWSME, that bulges at low-temperature and flattens at high, were obtained. Structural tests to show the differences between the deflections at the center of the bump caused by two different training methods were carried out afterwards.

**Introduction**

As one kind of smart materials, SMA has played a significant role in the application for smart structure systems. It owns the ability to change its macroscopic shape via a temperature-induced phase transformation between two phases. Applications for morphing aircraft based on SMAs are of great scientific interest in nowadays (Barbarino et al., 2014). Inspired by the mainly used SMAs for twist and camber morphing(Iii et al., 2007), this work addresses an adaptive 3-D TWSME bump structure with clamped boundaries.

**Preparations and Training Methods**

It is well known that the memory of both a hot shape and a cold shape is not a natural feature of the alloy, but rather is a 'learned' or 'trained' behavior. Researchers have made efforts to explore the different training methods for obtaining the TWSME in laboratory. While training methods for wires (Luo and Abel, 2007) and plate shaped SMA structures (Hartl, Lagoudas, et al., 2010) have been studied for many times, the approaches for 3-D structures are rarely mentioned. In this paper, two typical training methods that appear good effects on obtaining the TWSME are used and summarized in table 1. The deformation area of the bump shaped SMA is a disk with a radius of 40 mm and the original material is kept at 450°C for 30 minutes with furnace cooling afterwards. Each training progress took more than 20 thermal cycles.

**Table 1**. Two different training methods

| Training type | Training procedure |
|---|---|
| Constrained cycling of deformed austenite | 1. Cool the specimen below Mf<br>2. Load it in martensite state to a desired cold shape<br>3. Heat and cool it in the constrained condition through Mf and Af<br>4. Repeat step 3 a number of times |
| Combined shape memory and pseudoelastic cycling | 1. Heat the specimen above Af and below Md<br>2. Load it in austenite state to a desired cold shape<br>3. Cool it in the loaded condition below Mf<br>4. Unload it completely<br>5. Repeat the above steps a number of times |

**Results and Discussions**

Structural tests are carried out after the training process. Two different results are shown in figure 1 and figure 2, respectively. The whole temperature range of the test process is from about -100°C to 100°C, and the mid deflection of the bump structure is measure by the laser displacement sensor. 6 testing cycles were taken to obtain the stable results.

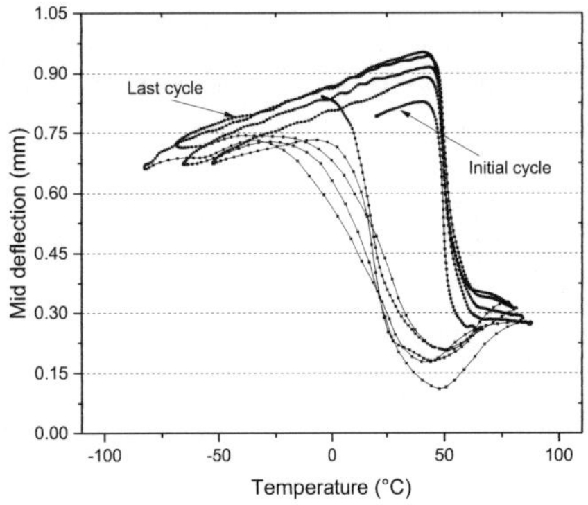


**Figure 1**. Displacement and temperature curves of SMA bump with constrained cycling of deformed austenite

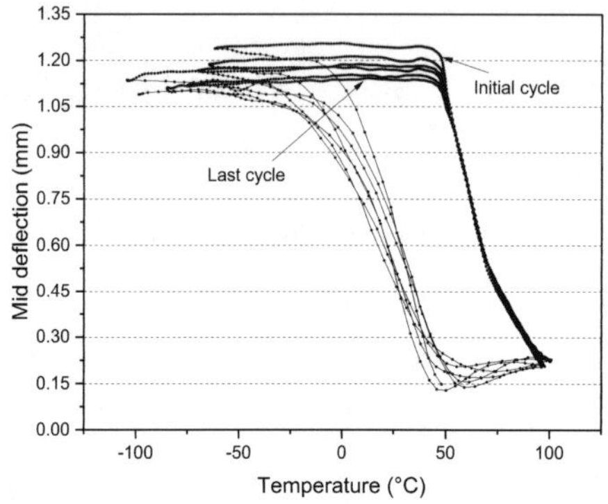


**Figure 2**. Displacement and temperature curves of SMA bump with combined shape memory and pseudoelastic cycling

From the figures showed above, obvious differences can be observed. The maximum deflection of the two SMA bumps with different training methods is about 0.8 mm and 1.1 mm, respectively. Also, a significant thermal expansion and contraction phenomenon occurs during heating and cooling stages in figure 1. Moreover, combined shape memory and pseudoelastic cycling training method appears a better TWSME than the first one. This phenomenon can be explained that the effective dislocation arrangements can mainly be introduced during the forward transformation by inducing the newly nucleated martensite into the particular martensite that is oriented to the applied stress. These dislocation arrangements give rise to the formation of the cold shape upon cooling with little influence on the recovery of the hot shape upon heating. In addition, the repeated loading progresses while training give the material a more stable bump-like shape which minimize the thermal expansion and contraction during the deformation.

In this research, differences between two training methods for 3-D SMA bumps are discussed. However, applying this kind of TWSME structure to morphing structures or actuators may not yet be mature enough, and more training details and low-cycle fatigue studies will be carried on in future works.

## Acknowledgment

This research was supported by National Natural Science Foundation of China (No. 11372133) and the Fundamental Research Funds for the Central Universities (No. NE2015001).

## References

*Barbarino, S. et al. (2014) 'A review on shape memory alloys with applications to morphing aircraft', Smart Materials and Structures. {IOP} Publishing, 23(6), p. 063001.*

*Hartl, D. J. et al. (2010) 'Use of a Ni60Ti shape memory alloy for active jet engine chevron application: I. Thermomechanical characterization', Smart Materials and Structures, 19(1), p. 015020.*

*Iii, R. D. V. et al. (2007) 'One Dimensional Morphing Structures for Advanced Aircraft'.*

*Luo, H. Y. and Abel, E. W. (2007) 'A comparison of methods for the training of NiTi two-way shape memory alloy', Smart Materials and Structures, 16(6), pp. 2543–2549.*

**Gap in pagination due to unavailable papers.**

**Pages 51-52**

# Damping Characteristics of SCSMA Vibration Isolator

**Kasumi Hayashi*, Shimpei Sato, Moe Kawamura, Nobuhisa Katsumata, and Ken Higuchi**

*Muroran Institute of Technology, Muroran, Hokkaido, Japan*

*19042063@mmm.muroran-it.ac.jp

**Abstract**

The vibration and/or impact isolator is proposed by using Single Cristal Shape Memory Alloys: SCSMA. The damping performance comes from the SCSMA's hysteresis characteristics of stress-stain curve. In this study, the vibration isolators, whose whole components consist of metal materials, are developed for applying to high temperature circumstance, and their damping characteristics are evaluated by vibration tests. Through the sweep test of constant acceleration, the damping performance dependency on the amplitude is investigated and the non-linearity of vibration response is confirmed.

## Introduction

The vibration isolator devices are used in wide fields, such as aseismic equipment, automobile suspension, and so on. In the field of aerospace, a vibration damper for engine mount, even in high temperature circumstances, is required to protect the structure and the onboard components from the fatigue. In addition, the planetary landers require the shock absorbing legs to protect the mounted equipment from landing impact. Furthermore, the reusable impact absorbing legs or devices will be required in a near future. In this study, the damping performance of the SCSMA vibration isolator is evaluated by the sweep and the random vibration tests.

## Merits of SCSMA

One of special features of SMA is the super-elasticity over the transformation temperature. When vibration is applied to a super-elastic material, the stress repeats loading and unloading, and the stress cycle due to the hysteresis characteristics of the stress-strain curve absorbs a vibration energy, as shown in Figure 1. Therefore, the damping effect can be obtained by the super-elastic material, with appropriate stress cycle condition.

Comparing with Ni-Ti based SMA materials, the Cu-Al-Ni based SCSMA (e.g., TiNi Aerospace Web) have the following features:

- Low stress value for martensitic transformation
- Narrow stress range in the hysteresis loop
- High strain recovery rate: 9%

The authors think that the above three features have advantages to design vibration isolators. The configuration of SCSMA damper developed as a prototype is shown in Figure 2. The set screws, shown in right figure in Figure 2, can apply the pre-strain to the SCSMA pins by 5%. In this case, the stress condition of SCSMA pins is located at the center of the hysteresis curve, as shown blue area in Figure 1, and each SCSMA pin radially located works as damper during vibration.

## Vibration Test

To investigate the damping performance of the developed SCSMA damper, the frequency sweep test of constant acceleration and the random excitation test are examined. Table 1 shows the test conditions. To investigate the relation between the vibration amplitude and the damping performance, the sweep tests are examined by three types of total mass and four types of vibration level, as shown in Table 1.

## Results and Discussions

Figure 3 shows the response curves of sweep tests under the four vibration levels, and Figure 4 shows the response curves of sweep tests under three total mass condition. Through these results, the differences of input acceleration to SCSMA pins affect the acceleration response magnification, and the response acceleration rate affect the response amplitude of SCSMA pins. When the response amplitude of SCSMA pins increase, the shape of hysteresis loop, furthermore the effective area of hysteresis curve for damping is also changes and several damping performances are obtained, as shown in Figure 5. This amplitude dependency of damping performance is important for designing the SCSMA damper, and an appropriate design of SCSMA pins, which controls the stress or strain value of SCSMA pins under vibration, is the key to obtain the effective damping performance. In addition, comparing the sweep up with the sweep down response curves, a jumping phenomenon around the peak frequency is observed as a non-linearity of vibration response.

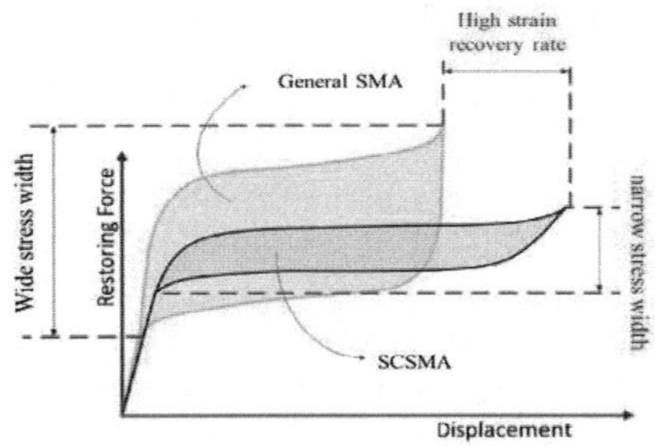


**Figure 1**. Hysteresis loop of SMA and SCSMA.

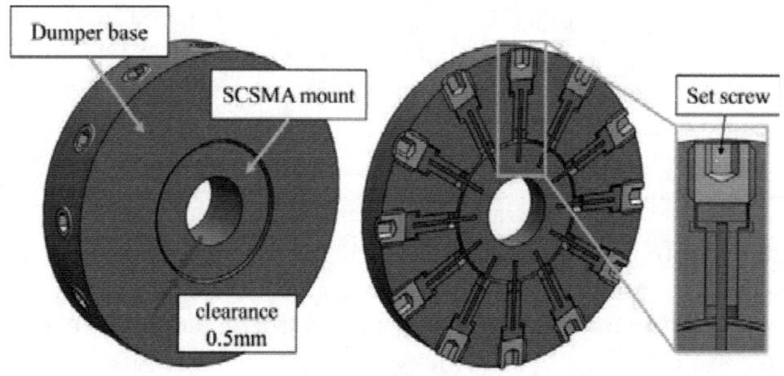


**Figure 2**. Schematic pictures of SCSMA damper.

**Table 1**. Vibration test configurations

| Total Mass (kg) | Vibration level (G) | Frequency range (Hz) | Sweep direction |
|---|---|---|---|
| 0.24, 0.34, 0.44 | 0.5, 1.0, 2.0, 4.0 | 100 ~ 1500 | One way: Up or Down |

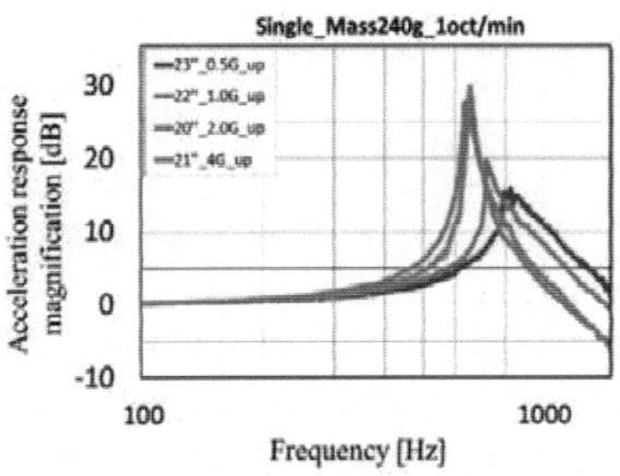


**Figure 3**. Response to the change in input acceleration.

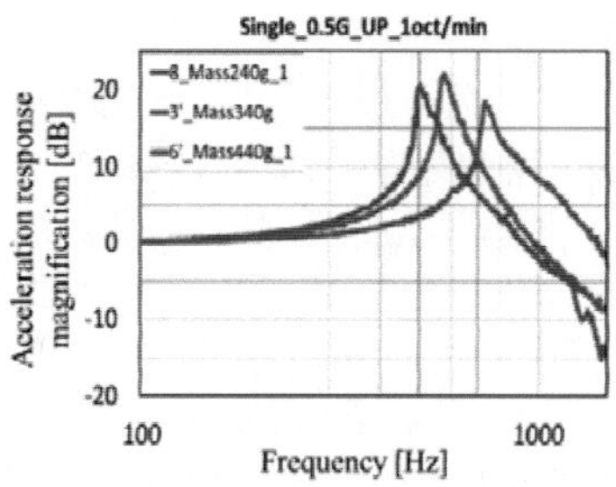


**Figure 4**. Response to the change in mass.

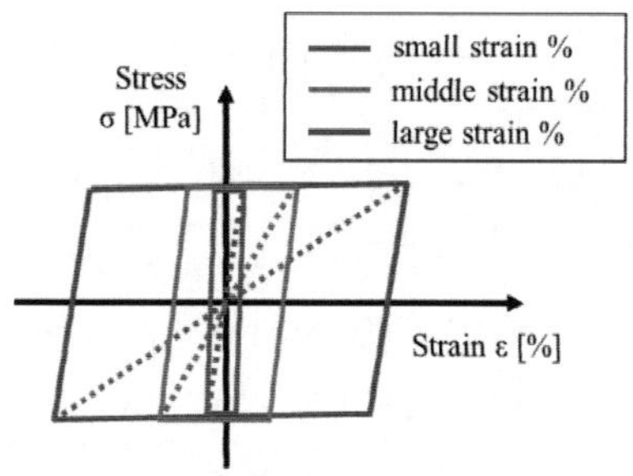


**Figure 5**. Schematics of the damping performance dependency on the amplitude.

## References

*TiNi Aerospace web site: https://tiniaerospace.com/products/scsma-components/*

30th International Conference on Adaptive Structures and Technologies
ICAST

# Runnability Improvement of the Moon Rover with Leg-Circle Transformable Wheel

**Ayako Torisaka*, Kohei Eguchi, Satoshi Miura, Victor Parque, and Tomoyuki Miyashita**

*Dept. of Aerospace and Astronautics, Tokyo Metropolitan University, 6-6 Asahigaoka Hino, Tokyo, 191-0065, Japan*

**ayakot@tmu.ac.jp*

**Abstract**

In this paper, we proposed a wheel equipped rover with deployable legs that can change the apparent wheel radius in order to improve the runnability of the rover which travels on the ground covered with lunar regolith. Furthermore, we formulated the driving force of the wheel based on Terramechanics and clarified the mechanism of the driving force change using our proposed rover wheel. In addition, the driving force was compared between the original wheel configuration and the leg expanded one in single wheel running experiments, and it was validated that the proposed leg expandable system exhibits higher driving force than the original circular wheel. With this system, in the case of a flat ground surface that does not require a high driving force, the vehicle should use the original wheel state, and when a high driving force is required such like the situation the vehicle need to escape the wheel from the local concaved ground, the proposed expandable leg system can be useful, and then the possibility of traveling on irregular surface efficiently not only on the moon but also on other planets was shown.

**Introduction**

One of the unmanned moon exploration techniques is a rover use that autonomously moves the moon surface. The rover has to move and explore terrain that has slope of 25 [deg] or more and is covered with soft slippery sand called regolith. However, in such soft ground, a slope of 20 [deg] is said to be the climbing limit of rigid wheels, and it is an issue to improve the running performance of the rover. When going uphill, a towing load is generated on the slope in the direction of the slope, and when the sum of the driving force of the front and rear wheels is larger than this, the rover can go uphill. In addition, the tow load also increases with the increase of slope. Here, while the wheel increases the driving force by slip, eventually the towing load becomes larger than the driving force and reaches the climbing limit. Therefore, improvement of the driving force is considered to be an important factor to improve the runnability of the lunar explorer.

**Methods/Model Formulations**

In this study, the driving force is evaluated using Teramechanics. Terramechanics is the dynamics of rover driving targeting its wheels and the ground surface. The amount of wheel sinking is calculated using the parameters of wheel shape, running condition, and regolith property. Then, the incidence angle of the wheel with respect to the ground and the escape angle are obtained from the obtained amount of wheel sinking. Next, the contact area of the wheel is determined from the incidence angle and the escape angle, and the vertical stress from the ground to the wheel and the shear stress occurring there are calculated. Finally, the driving force $F_x$ is calculated by integrating the vertical stress and the shear stress in the wheel traveling direction. $F_x$ is formulated as equation (1) using the wheel radius r, the wheel width b, the incident angle $\theta_f$, the escape angle $\theta_r$, the vertical stress $\sigma(\theta)$ and the shear stress $\tau(\theta)$ (Wong,1967).

$$F_x = rb\int_{-\theta_r}^{\theta_f}\{\tau(\theta)\cos\theta - \sigma(\theta)\sin\theta\}d\theta \quad (1)$$

Since the driving force is proportional to the wheel radius r, this research targets operations requiring a high driving force, as represented at the time of climbing a hill, we propose a deployable wheel that can obtain higher driving power by equipping the deployable leg to earn the wheel radius r as shown in Figure 1.

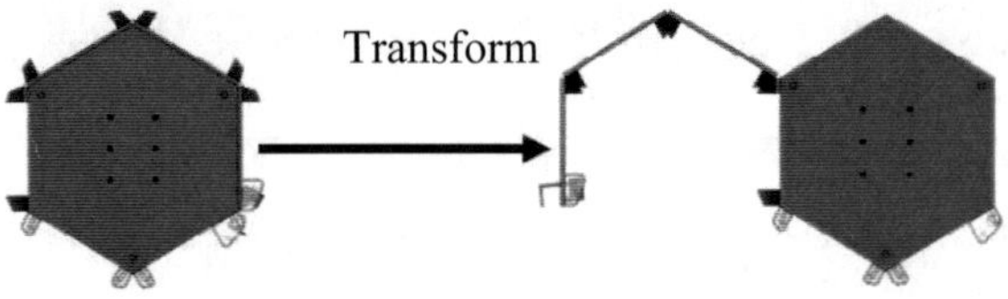


**Figure 1**. The leg-circle transformable wheel

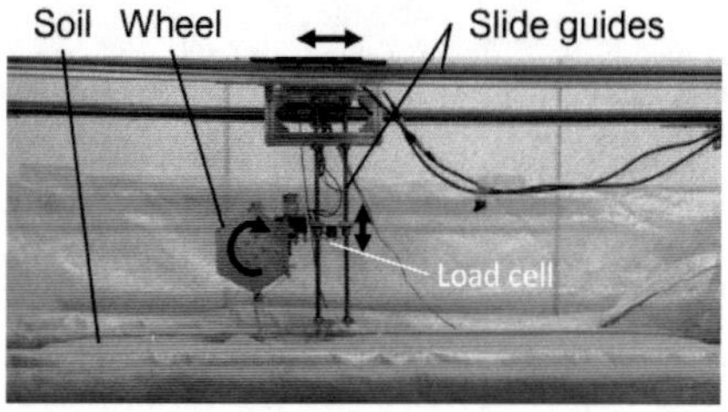


**Figure 2**. Overview of the single wheel test bed

**Experiments and Simulation**

In order to measure the driving force, a single wheel experiment on regolith is carried out using a model shown in Figure 2. The wheels are driven by a motor and can be moved vertically and horizontally by slide guides. In addition, the driving force is measured by a force sensor, and the moving distance and the wheel rotation angle are measured by rotary encoders. Then, in order to clarify the mechanism of driving force increase by the proposed wheel, driving force simulation based on Teramechanics is carried out and is compared with the experiment results.

**Results and Discussions**

From the experimental results, it was found that the proposed expandable leg system shows about 1.5 times the driving force as compared with a general wheel state (Figure 3). Also, the wheel usually moves forward with the friction between the ground as the driving force. On the other hand, in the case of the expanded leg state, it is clarified that the effect of adding the motor torque to the driving force generate large effect on obtaining higher driving force than the general wheel state. (Figure 4).

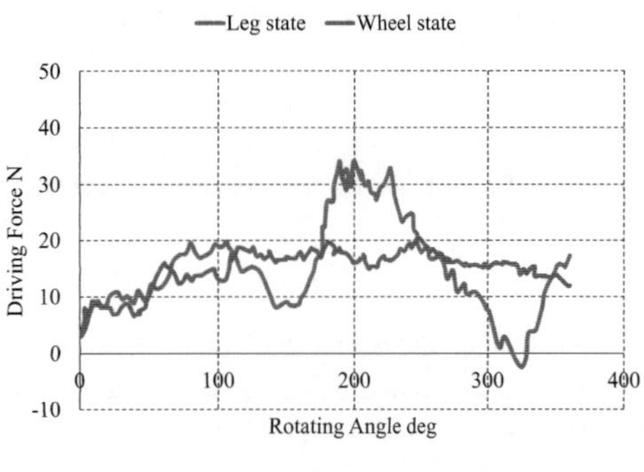


**Figure 3**. Comparison of the driving force

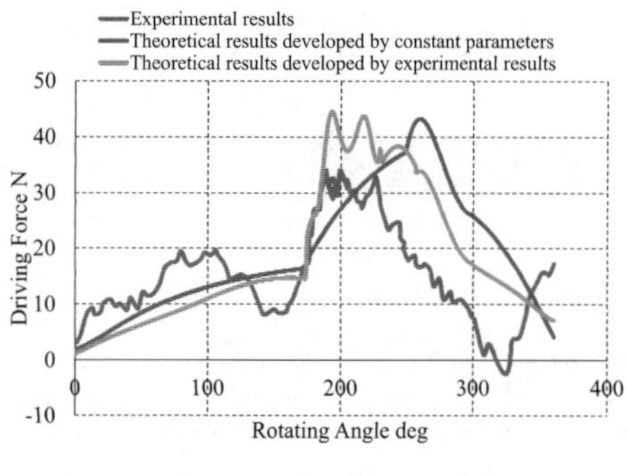


**Figure 4**. Experimental results and theoretical results

**References**

*Wong, J. Y., Reece, A. R., (1967) Prediction of Rigid Wheel Performance Based on the Analysis of Soil-Wheel Stresses Part I. Performance of Driven Rigid Wheels, Journal of Terramechanics, vol. 4, No. 1, pp. 83*

# Incorporating Sensing Capability in an Electrorheological Haptic Module

**Alex Mazursky[1], Tae-Heon Yang[2], Sam-Yong Woo[3], and Jeong-Hoi Koo[1*]**

[1]*Department of Mechanical and Manufacturing Engineering, Miami University, Oxford, OH, 45056, USA*

[2]*Department of Electronic Engineering, Korea National University of Transportation, Chungju-si, Chungbuk, 27469, Republic of Korea*

[3]*Center for Mechanical Metrology, Korea Research Institute of Standards and Science, Daejeon, 34113, Republic of Korea*

[*]*koo@miamioh.edu*

**Abstract**

We present embedding a sensing capability to a slim haptic actuator based on electrorheological (ER) fluids, designed for conveying vivid kinesthetic and tactile sensations at small scale. Haptic feedback is produced through electrorheological fluid's controllable resistive force and varies with the actuator's deformation. To demonstrate the proposed actuator's feedback in realistic applications, a method for measuring the actuator's deformation must be implemented for feedback control. To this end, in this study, we incorporate a sensor design based on stress-sensitive resistive film in bending to the ER haptic actuator. The combined actuator and sensor module was tested for its ability to simultaneously actuate and sense the actuator's state under indentation. The results show that the deflection sensor can accurately track the actuator's displacement over its small stroke range. Thus, the proposed sensor may enable control of the output resistive force according to displacement.

## Introduction

Haptic technology has become essential in electronic devices today to bridge the gap between our physical world and virtual information through the sense of touch. Haptic feedback consists of two sensations: kinesthetic and tactile. Kinesthetic feedback is acquired by receptors in joints, muscles and ligaments, and provides information about force and displacement. Tactile feedback is observed by cutaneous receptors and provides information about vibration, texture and friction (Srinivasan and Basdogan, 1997). For the greatest immersion when designing for haptic feedback, both modes of feedback must be present.

The authors developed a haptic actuator based on electrorheological fluid that can convey both kinesthetic and vibrotactile sensations in a compact form (Mazursky et al., 2018, 2019). To demonstrate this actuator in real-world applications, such as haptic rendering of compliant objects, a method of sensing the actuator's state is necessary to incorporate control. Thus, the primary goal of this study is to embed a displacement sensor within a haptic button actuator based on mixed modes (squeeze and flow) of ER fluid, where sensations are generated from fluid flow through charged electrodes due to the user's interaction. Hence, to control the actuator's force output with respect to the user's finger position, this study investigates the addition of a stress-sensitive sensor to measure the displacement of the actuator's contact surface.

## Methods

The mixed mode ER haptic actuator with the proposed sensor are presented in Fig. 1. When a user presses down on the contact surface, ER fluid beneath it is driven through squeeze and valve electrodes. Therefore, the resistive force felt in the user's finger directly corresponds to the fluid's yield stress under applied field. For a range of applied voltage magnitudes and frequencies, a range of haptic sensations may be felt by the user. The deflection sensor consists of a stress-sensitive film with resistance sensitive to bending. The sensor is adhered to the surface of the haptic actuator and spans halfway across the contact surface. Thus, when the user interacts with the device, they press directly on the bending sensor and contact surface. By measuring the change in voltage across the sensor, the indentation depth may be approximated in real time.

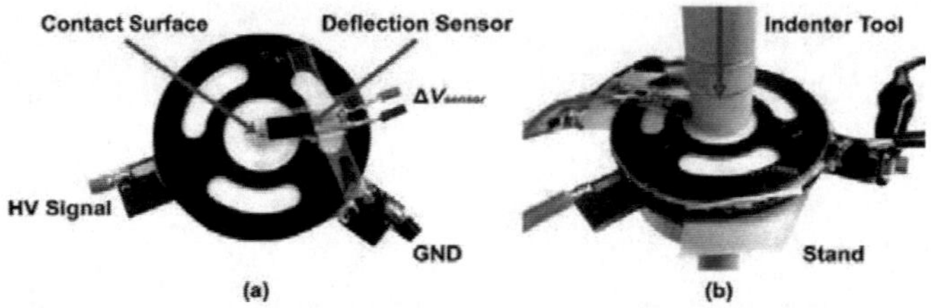


**Figure 1**. **(a)** Top-down view of the combined haptic actuator-sensor system and **(b)** the experimental setup.

To test the simultaneous performance of the actuator and sensor, mechanical analysis was conducted with a dynamic mechanical analyzer. As shown in Fig. 1b, this experimentation measured the total resistive force with respect to indentation depth over the actuator's 1 mm stroke. Two cycles of loading/unloading were performed at a rate of 1 mm/s. A function generator and voltage amplifier were used to supply the high voltage (HV) to the actuator's electrodes. A microcontroller was used to record the voltage across the sensor over time.

## Results and Discussion

Figure 2 presents the combined actuator (force vs. time) and processed sensor (indented depth vs. time) data from the described experiment. Figure 2a and b show the response when 0 kV and a 4 kV 3 Hz sine wave are applied, respectively.

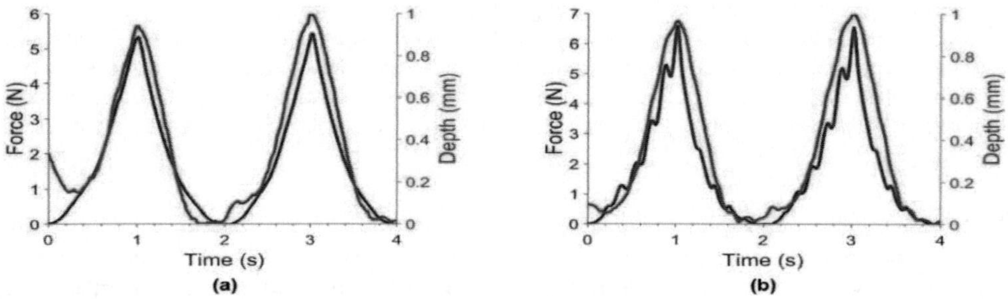


**Figure 2**. Actuator and sensor response subjected to **(a)** 0 kV and **(b)** 4 kV sine wave at 3 Hz.

As shown, the deflection sensor sufficiently tracks the actuator's displacement (1 mm over 1 second) and corresponding force output for both the off-state and high voltage. Therefore, the proposed sensor can enable control of the output resistive force according to displacement to demonstrate realistic applications of the ER fluid-based haptic module. Furthermore, the sensor does not impede upon the tactile and kinesthetic sensations (observed as oscillations in force in Fig. 2b) and the actuated forces do not influence the sensor's response.

## References


*Mazursky A, Koo J-H and Yang T-H (2018) Experimental evaluation of a miniature haptic actuator based on electrorheological fluids. In: SPIE Smart Structures and Nondestructive Evaluation, Denver, CO, USA, 2018. SPIE.*

*Mazursky A, Koo J-H and Yang T-H (2019) Design, modeling, and evaluation of a slim haptic actuator based on electrorheological fluid. Journal of Intelligent Material Systems and Structures: 1–13.*

*Srinivasan MA and Basdogan C (1997) Haptics in virtual environments: Taxonomy, research status, and challenges. Computers & Graphics 21(4): 393–404.*

# Development of PZT Fiber Sensor Array to Locate Acoustic Source

**Bao Ha and Gang Wang***

*Department of Mechanical and Aerospace Engineering*

*University of Alabama in Huntsville, Huntsville AL USA 35899*

**gang.wang@uah.edu*

## Introduction

Because of their electro-mechanical coupling property, piezoelectric materials have been widely used for sensing and actuation applications. Piezoelectric materials have been used as surface-mount sensors to measure physical movement or strain by directly bonding or embedding piezoelectric thin sheets onto the structure (Chang, 1998; Giurgiutiu, 2007). Piezoelectric ceramics are perhaps the most popular piezoelectric materials for sensing purpose. In particular, sensors made of piezoelectric ceramic patches have recently gained increasing popularities in the field of structural health monitoring (SHM). For example, recent research works on Lead-Zirconate-Titanate (PZT) sensor has made it a quite promising technique for nondestructive evaluation (NDE) applications. Recent advancements in this area include applications of guided ultrasonic Lamb waves for identification of damage in composite. The Acousto-ultrasonic (AU) approach has been used to identify damages in engineering structures (Worlton, 1957; Viktorov, 1967; Rose, 2002; Su and Ye, 2009).

In this paper, a PZT fiber (i.e., d33 effect) sensor array is proposed to conduct Lamb wave based damage detection in panel-like structures by exploring its best directional sensing capability. Simple interrogation schemes were developed as well to effectively determine Lamb wave propagation characteristics and acoustic source location. Comprehensive tests were conducted to validate the proposed damage detection concept and evaluate the performance of the PZT fiber array in a plate structure. We expect to extend the same concept for the wave based SHM demonstration and applications.

The novelty of the present effort includes: (1) tailored sensor array design using PZT fiber for the best directional sensing; 2) in-house fabrication of PZT fiber based array; and (3) comprehensive test to determine Lamb wave propagation characteristics at different actuation frequencies; and 4) simple interrogation scheme to locate acoustic source.

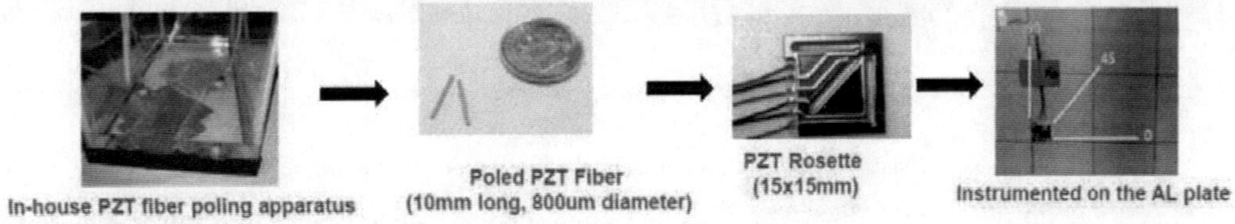


**Figure 1.** Fabrication of a PZT fiber sensor array

## Methods

Figure 1 shows the process for the proposed PZT fiber array fabrication. First, we developed a PZT fiber poling apparatus and successfully poled PZT fibers. Then, we designed and fabricated PZT rosette prototype. Finally, the proposed PZT rosette was surfaced mounted on a plate structure for Lamb wave sensing. Figure 2 shows the collected Lamb wave signals, in which a pair of circular PZT disks was used to actuate the plate structure from 20 kHz to 100 kHz. Based on these collected wave propagation signals along three different directions, the goal is to locate acoustic source (i.e., actuation location). Collected voltage signals in each PZT fiber are related to the strain measurements.

## Results

We simply computed the principal strain direction using three sensor results under each frequency. Associated principal strain angle represents the angle between actuator and sensor array. Then, we are able to locate acoustic source. Figure 3 shows the acoustic source location results. There is a discrepancy between actual and calculated angle. However, the error is within 15%. Note that we could not ensure a perfect alignment for a PZT array when it is surface bonded. This could contribute to the error.

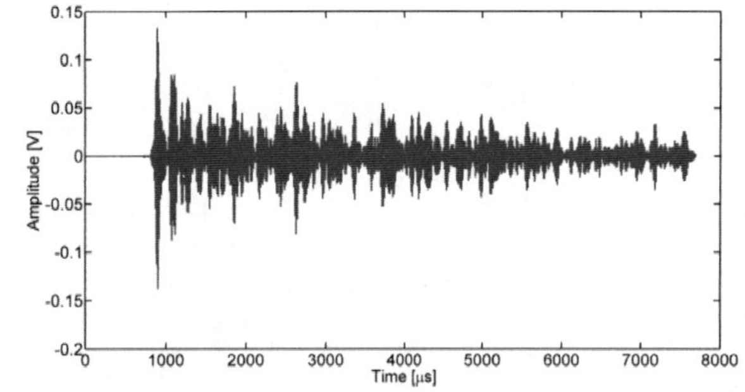


**Figure 2.** Collected PZT fiber signals

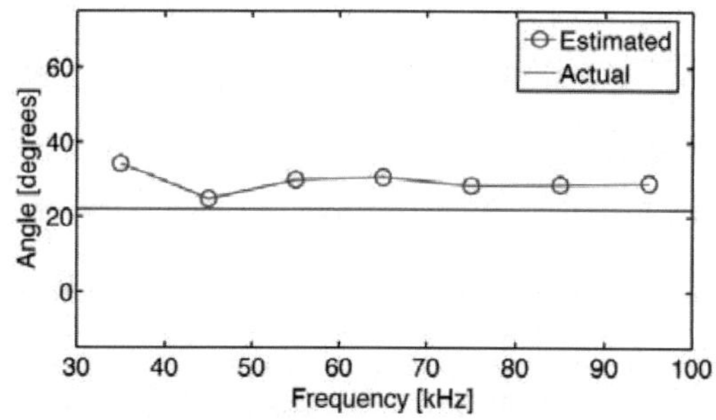


**Figure 3.** Acoustic source location result

In summary, we successfully designed and prototyped the proposed PZT fiber sensor array. In addition, we demonstrated the Lamb wave sensing capabilities of such array. Finally, we are able to locate acoustic source via a simply computation. Similar approach can be extended to the damage case. Associated signal processing and interrogation scheme must be carefully designed.

## References

*Chang, Fu-Kuo, eds, Structural Health Monitoring: Current Status and Perspectives, CRC Press, Boca Raton, FL, USA, 1998.*

*Giurgiutiu, V. Structural Health Monitoring with Piezoelectric Wafer Active Sensors; Academic Press: Burlington, MA, USA, 2007.*

*Rose, J. L., "A baseline and vision of ultrasonic guided wave inspection potential," Journal of Pressure Vessel Technology, Vol. 124, No., 3, pp. 273–282, 2002.*

*Su, Z. and Ye, L. Identification of Damage Using Lamb Waves: From Fundamentals to Applications, Springer Verlag, London, UK, 2009.*

*Viktorov, I. A., Rayleigh and Lamb Waves—Physical Theory and Applications, Plenum Press, New York, NY, USA, 1967.*

*Worlton, D. C., "Ultrasonic testing with Lamb waves," Nondestructive Testing, Vol.15, pp. 218–222, 1957.*

# Smart Elastomers for the Sensing of Force and Vibration: A Proof of Concept

**Jorge de Jesús Lozoya-Santos[1*], Adriana Vargas Martínez[1], Juan Carlos Tudón-Martinez[2], Ruben Morales-Menendez[1], Ricardo A. Ramirez-Mendoza[1], and Armando Roman-Flores[1]**

[1]*School of Engineering and Sciences, Tecnologico de Monterrey, Mexico*

[2]*School of Engineering and Technology, Universidad de Monterrey, Mexico*

[*]*jorge.lozoya@tec.mx*

**Abstract**

A smart sensor is proposed for the measurement of vibration and forces on automotive structures. The concept includes transducer, energy harvesting, energy storage and wireless transmitter. The use of smart materials consisting of smart elastomers using carbon nanotubes and piezoelectric sensors shows potential applications in chassis measurements. All the generated information from the sensor is recorded in the cloud on real time.

**Introduction**

The importance to develop a smart sensor that is involved in the automotive industry is of great importance, since this industry is quickly evolving and creating new technologies that improve the quality of performance for the user. It's crucial to detect any kind of damage in any structure before it is too late, it helps the user avoid primarily accidents and costs, Nagayama, T. and Spencer Jr, B.F., 2007. Specifically talking about chassis, the main issue is the expensive technology and high specialization of personnel is required in order to perform the structural analysis. Installing a Smart Sensor in chassis would enable automotive companies to perform field tests and get real time information for its future analysis and the part re-designing process, Kumbhar, S.R. and SubhasisMaji, B.K., 2013.

A smart sensor is distinguished because it is able to join four systems: transducer, energy harvesting, energy storage and wireless transmitter. This work study how a MagnetoRheologic Elastomer (MRE) and a Piezoelectric-based structure can substitute a strain gauge when used as a transducer that it is able to measure the motion from the variation of an electric resistance and capable of generating electric energy, when its placed on a metallic surface while being submitted to different vertical motion and force loads.

**Methods/Model Formulations**

For the harvesting of energy, an Energy Harvesting System is proposed using a set of piezoelectric sensors placed within a rubber plate, embedded into a support inside the case designed for this system, complementing it with a circuit designed for the storage of the energy generated. A proof of concept is presented to obtain the data. It is produced by the arrangement. Then, a series of experiments are conducted consisting of the rubber plates being attached to a system that is able to produce mechanical vibrations in a range of frequencies, from 4 to 10 Hz, while at the same time changing the length of the rubber plate from 8 cm to 5.5 cm. Regarding to the MREs, these are tried as sensors for the purpose of measuring the deformation of a metallic structure, while at the same time monitoring the strength of the material at the same time that its shape is changing. A prototype is proposed. A proof of concept, including the design and implementation of four MREs laminates with different patterns, evidences, through experiments such as tension and compression tests, a potential linear behavior of electric resistance versus applied force. Subsequently, the signals emitted by the voltage acquired by the sensor go through a processing unit (Arduino UNO), and then are transmitted wirelessly, using Wi-Fi as a communication protocol, reaching a server (cloud) for the storage of the information collected by the sensor.

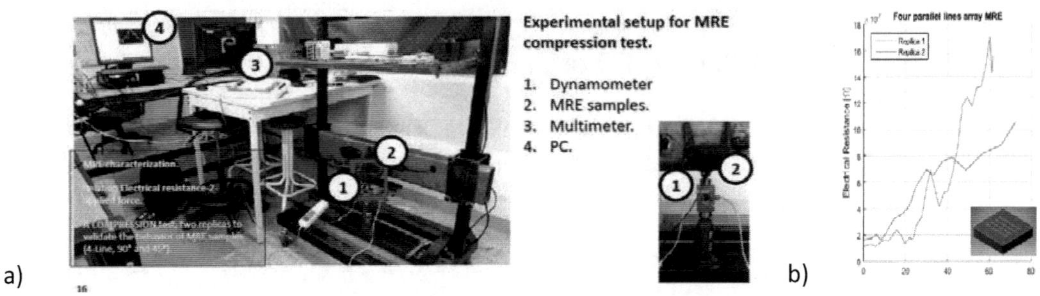


**Figure 1.** Proof of concept for the MRE sensor: a) compression test using a chassis fragment, b) electric resistance versus applied force.

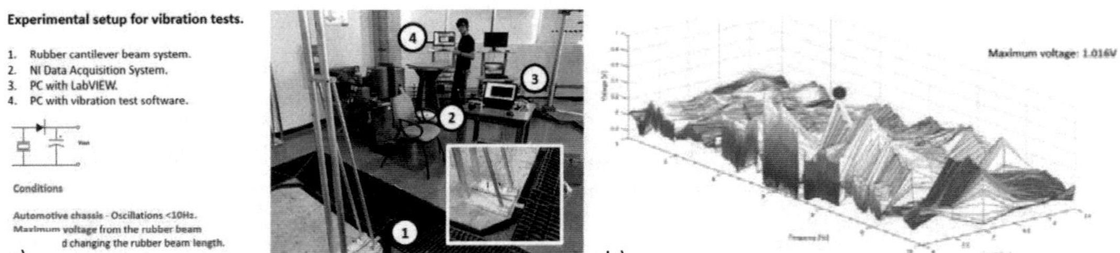


**Figure 2**. Proof of concept for the energy harvesting device: a) Vertical vibration test for the energy harvesting mechanism, b) generated voltage vs rubber beam length vs vertical frequency.

## Results and Discussions

The MRE proof of concept, shows that through the pattern of the particles in the elastomer is possible to get a linear relation in compression and tension test between force and electric resistance, Fig. 1. Regarding the energy harvesting device, the ideal length of the rubber plate should be 7.5cm and the ideal frequency should be at 7 Hz. With this specification the piezoelectric will produce the most voltage comparing all the voltages with all the lengths, from 4 Hz to 10 Hz of every rubber plate tested, Fig 2. An approach for an alternative to the strain gauge was created: MR elastomer, which responded with an electrical resistance variation when a force was applied on it. The energy harvesting system can generate voltage, at the optimal length of 7.5 cm during a frequency of 7 Hz for the current configuration. The dashboard is able to show the information obtained by the Google spread sheet, it takes roughly a minute to refresh the data. Lastly, the smart sensor will collect information of the current state of an automotive chassis while displaying the obtained data in real time, at the same time also implementing the equivalent method to be able to post analysis computer simulations and be able to constantly improve the design.

## References

*Nagayama, T. and Spencer Jr, B.F., 2007. Structural health monitoring using smart sensors. Newmark Structural Engineering Laboratory. University of Illinois at Urbana-Champaign.*

*Kumbhar, S.R. and SubhasisMaji, B.K., 2013. Development and testing of MRE bushing for road vehicle suspension system. International Journal of Recent Advances in Mechanical Engineering (IJMECH), 2(3), pp.37-44.*

# An Optical RGB Sensor for Monitoring the Degradation of Magnetorheological Fluids in High-torque Clutch Actuators

**Benoit Gillet*, Mathieu Lamy, Alexandre Landry-Blais, Raphaël Pilon, and Jean-Sébastien Plante**

*Faculté de génie, Université de Sherbrooke, Centre 3IT, 3000 boulevard de l'université, Sherbrooke QC J1K0A5 Canada*

**benoit.gillet@usherbrooke.ca*

## Introduction

The aerospace industry is gradually moving toward "more electric aircraft" flight controls having the potential to improve functionality while having lower environmental footprints compared to conventional hydraulic systems. To this end, an electrical flight control system using continuously slipping magnetorheological (MR) clutches and electric motors has been demonstrated to have superior dynamic performance compared to hydraulics while being lightweight and capable of meeting stringent reliability requirements for primary flight control applications (Chouinard, 2014). A key challenge that must be addressed for a wide spread deployment of the MR technology in aerospace is the ability to monitor MR fluid condition while it ages in operation.

MR fluid aging is affected by two interconnected phenomena: in-use thickening and particles oxidation. Hard particles tend to flake their oxidized surface into smaller particles (Iyengar, 2003), while soft particles tend to deform and agglomerate, in both cases, resulting in a thicker fluid. Moreover, particles oxidation can decrease the maximum transmissible shear stress. (Sunkara et al., 2009).

MR fluid condition can be monitored by periodically measuring the clutch open-loop torque to current response. With this technique, a clutch is considered having reached its end-of-life when the transmitted torque of the clutch decreases by 30% compared to its initial value. Since flight control is a vital system, another completely independent mean of measuring fluid condition is mandatory to ensure reliable fluid monitoring. There is therefore a need to develop such a dedicated sensor for monitoring MR fluid condition throughout aging, and to date, no such efforts have been reported in the literature.

## Sensor Design and Integration

A survey of relevant physical principles that could be used to monitor MR fluid condition has been made considering electric, magnetic, optical, and acoustic properties. Basic experiments have been conducted on each of them and optical properties showed the highest sensitivity to degradation and has been selected as the working principle for sensor development.

The proposed sensor is built using a commercially available RGB board that measures the color and clearness of the fluid. The RGB board is placed in front of a clear window where it has an optical access to the fluid in operation. The RGB board and optical window are embedded in a 3D printed frame and shell for easy and rapid prototyping.

In order to avoid MR fluid stagnation in front of the RGB board, a continuous flow of MR fluid is ensured by using a self-pumping MR clutch design, based on a grooved drum configuration where shear surfaces act as a magnetic screw pump (Pilon et al., 2018).

## Experiment

The RGB sensor has been set up in a high-torque MR clutch designed for MR fluid degradation testing. The clutch is mounted on a fully instrumented durability test bench (Fig 1). The RGB sensor monitors the darkening of the fluid throughout aging. Sensor performance was studied with two fluids: a commercial fluid specifically designed for clutch operation, the LORD 140CG, and a homemade perfluoropolyether (PFPE) base oil fluid, the GPL-103.

As shown in Fig 2b, as the MR fluid ages, the clutch output torque slowly decreases for a constant value of current supplied to the coil. For a recirculating clutch design, fluid end-of-life is seen at ~27 MJ/ml where a sudden torque decrease is found. As shown in Fig 2a, the fluid darkening is measured throughout the test and a correlation between aging and darkening is observed.

Preliminary results with Lord 140CG has showed a darkening of 12% and 13% at the fluid end of life for two test conditions: 20 N.m at 150 RPM and 40 N.m at 100 RPM. Tests with GPL-103 fluid have shown a darkening of 21% for a test at 20 N.m at 150 RPM. Additional tests are currently ongoing for both fluids and will be added to the paper before the conference. Results show that an aging criterion based on a RGB sensor can be defined as a specific value of darkness that correlates with the 30% torque reduction of a specific formulation of MR fluid. If the darkening stays below this aging criterion, the fluid condition is considered to be good, otherwise the fluid has reached end of life. According to preliminary results, the criteria should be around 12% for the Lord 140CG and 20% for the GPL103.

Further applications of the sensor include remaining life estimation and MR fluid flow monitoring. Indeed, the RGB sensor can be used to predict the remaining lifetime of the fluid by estimating the average rate of darkening to estimate the remaining operational time before the darkness threshold is reached. Furthermore, the sensor is able to detect possible MR fluid flow blockage when the RGB sensor measures a local brightening of the fluid.

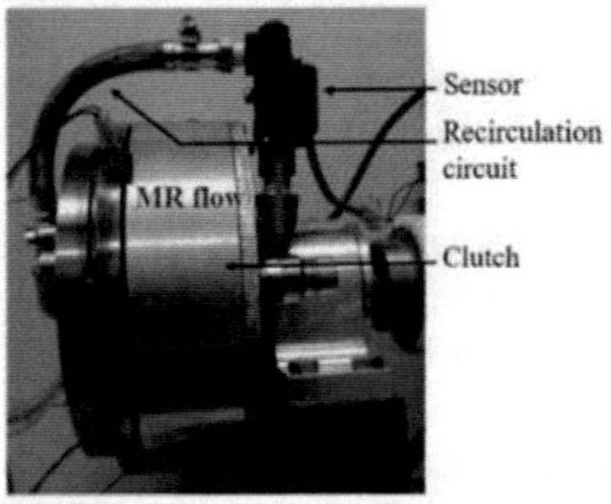


**Figure 1.** Clutch set up in the test bench

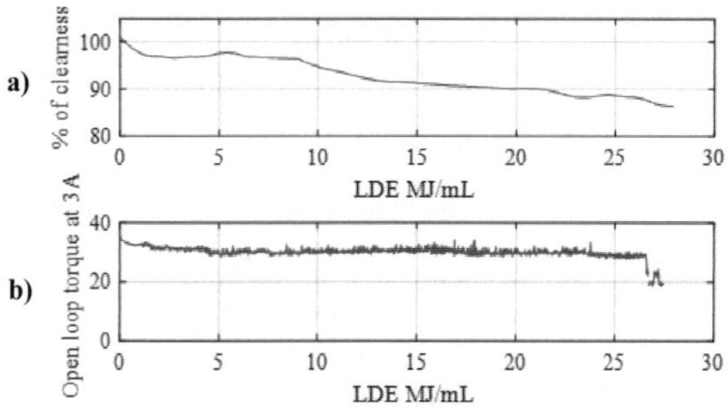


**Figure 2**. Sensor results for a Lord 140CG fluid. (a) Optical sensor correlation and (b) torque vs current relationship.

## Conclusion and Discussion

The proposed RGB sensor concept demonstrates the feasibility of using an optical sensor to monitor the degradation of MR fluid in a clutch device. According to preliminary results, the RGB sensor correlates with the degradation status of both fluids. The fluid is considered to be fully used when it reaches a pre-defined percentage of darkening corresponding to a loss of open-loop torque. Moreover, the sensor can estimate the remaining life of the fluid and detect when there is a no flow in the clutch recirculation circuit.

Future work are needed to further improve the sensor design as this is a proof of concept. These works should most certainly gather a wide body of experimental data on various MR clutch systems to provide statistically meaningful data to assess the accuracy, repeatability, and precision of the sensor. Moreover, studying various sensor integration scenarios on various MR clutch designs (e.g. disk, drums, with or without pump effects) is necessary to understand the effect of MR fluid circulation on sensor performance with MR clutch, with the limiting case of no-circulation.

## References

*Chouinard P (2014) Conception et validation expérimentale d'un système d'embrayages magnétorhéologique à glissement continu pour les commandes de vol d'aéronefs.*

*Iyengar VR (2003) Durable magnetorheological fluid compositions. US 6547983 B2.*

*Pilon R, Plante J-S, Charron F, et al (under review). A magnetic screw pump for durability enhancement of MR clutches.*

*Sunkara SR, Root TW, Ulicny JC, et al. (2009) Iron oxidation and its impact on MR behavior. Journal of Physics: Conference Series 149: 012081. DOI: 10.1088/1742-6596/149/1/012081.*

# Hilbert-Huang Transform on Piezoelectric-Piston Synthetic Jet

**Zhang Li***, **Li Hongsheng, Lu Hong, and Long Li**

*School of Automation, Nanjing Institute of Technology, Nanjing 211167, China*

*zdhxzli@njit.edu.cn

**Introduction**

In this paper, Hilbert-Huang transform (HHT) is used to deal with the nonlinear unsteady jet signal which generated by the synthetic jet driver. The piezoelectric-piston synthetic jet actuator is used to generate the synthetic jet. The instantaneous jet signal was obtained by the hot wire anemometer. Through the synthetic jet origin signal, 9 order Intrinsic Mode Functions (IMF) were obtained through the Empirical Mode Decomposition (EMD). When the IMFs as the basis functions, the Hilbert transform is applied for the original signal. The Hilbert spectrum was obtained. Through the analysis of the Hilbert spectrum, it's found that the double helix structure of DNA like graphics in the figure of high frequency resonance. And the frequency of the middle of the double helix line is just membrane vibration frequency of the synthetic jet actuator. This is the result of the resonance of the actuator cavity.

**Methods/Model Formulations**

The main content of HHT includes two parts. The first part is Empirical Mode Decomposition (EMD), and the second part is Hilbert Spectrum Analysis (HSA). Simply speaking, the basic process of HHT processing non-stationary signals is: firstly, the given signal is decomposed into several intrinsic mode functions (represented by Intrinsic Mode Function or IMF, also known as intrinsic mode functions), which are components satisfying certain conditions; secondly, Hilbert transform is applied to each IMF to obtain the corresponding Hilbert spectrum, i.e. each IMF is represented in the Hilbert spectrum. In the joint time-frequency domain; finally, the Hilbert spectrum of the original signal is obtained by summing up the Hilbert spectrum of all IMFs.

Finally, the original time-domain signal is analyzed by Hilbert spectrum based on IMFs, and the result is shown in Fig.1

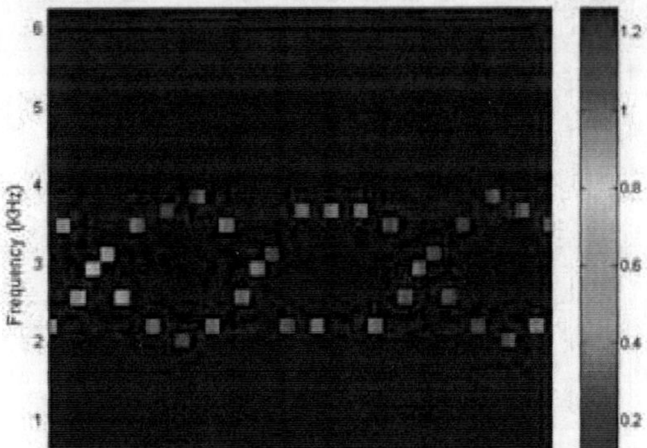

## Results and Discussions

Through the analysis of Hilbert spectrum. At the low frequency of the resonance signal, the jet signal is mainly generated by the vibration of the vibration film; at the high frequency of the resonance signal, the pattern similar to the double helix structure of DNA is found, and the midline of the double helix structure is exactly the excitation frequency value of the vibration film of the synthetic jet driver. This is the result of the resonance of the whole driver cavity.

## References


*[1] N. E. Huang, Z. Shen, and S. R. Long. The empirical mode decomposition and the Hilbert spectrum for non-linear and non-stationary time series analysis. Proc. R. Soc. London, Ser. A, 1998,454,903-995.*

*[2] N. E. Huang, Z. Shen, and S. R. Long. The mechanism for frequency downshift in nonlinear wave evolution. Adv. Appl. Mech., 1996,32,59-111.*

*[3] N. E. Huang and S. P. Shen. Hilbert-Huang Transform and its Applications. Interdisciplinary Mathematical Sciences-Vol 5, World Scientific Publishing, 2005.*

*[4] N. E. Huang, Normalized Hilbert transform and instan-taneous frequency. NASA Patent Pending, GSC, 2003, 14, 673-1.*

*[5] Crook A, Wood N J. Measurements and visualizations of synthetic jets[C]. American Institute of Aeronautics and Astronautics. 39th Aerospace Sciences Meeting and Exhibit, January 8-11,2001, Reno, Nevada.*

*[6] Zhong S, Garcillan L, Pokusevski Z, et al. A PIV study of synthetic jets with different orifice shape and orientation. American Institute of Aeronautics and Astronautics. 2nd AIAA Flow Control Conference, June 28-July 1, 2004, Portland, Oregon.*

# Development of a Multi-axis Active Seat Mount System for Helicopter Aircrew Whole-body Vibration Reduction

**Jason Chang, Amin Fereidooni, Viresh Wickramasinghe, and Yong Chen***

*Flight Research Laboratory, Aerospace Research Center, National Research Council Canada*

*1920 Research Road, Ottawa, Ontario, Canada, K1A0R6*

**Eric.Chen@nrc-cnrc.gc.ca*

**Abstract**

This paper presents the development and evaluation of a proof-of-concept multi-axis and actively controlled helicopter seat mount system for aircrew whole-body vibration reduction. The multi-axis seat mount is designed to be installed between the helicopter seat floor and the seat supporting structure to minimize the impact on crashworthiness requirements of the helicopter seat. The design involves multiple miniature force actuators to counteract the vibrations of the seat frame and occupant transmitted from the helicopter floor in three orthogonal directions. The actuators are controlled by an adaptive feedforward filtered-x least mean square algorithm to cancel the helicopter floor vibration input. The prototype active seat mount design was tested in various configurations with a shaker table providing representative Bell-412 helicopter vibration inputs. Test results demonstrated that the vibrations of the seat frame and mannequin occupant body were suppressed simultaneously, and the major N/rev harmonic peaks of the occupant's whole-body vibration were reduced by 95%. This demonstrated that the multi-axis active seat mount design can mitigate the whole-body vibration exposure of the helicopter aircrew to improve their ride quality and reduce adverse health effect.

**Introduction**

Helicopter aircrew are exposed to high levels of whole-body vibration during fight missions, which can adversely affect their ride comfort and perception in the short-term and may lead to adverse health issues such as lower back pain and neck strain in the long-term.

This paper presents the development and evaluation of a multi-axis active seat mount system to reduce the whole-body vibration exposure of the pilot. The active mount system involves the use of two parallel active mounts, each installed between the helicopter floor and the leg of the pilot seat frame to avoid adverse impacts on the crashworthiness of the helicopter seat. Each active seat mount is integrated with multiple miniature force actuators in three orthogonal directions to counteract the vibrations transmitted from the helicopter floor to the seat frame and occupant body. A multi-channel adaptive feedforward Filtered-x Least Mean Square (FxLMS) controller is developed to control the miniature actuators of the two active mounts simultaneously to actively cancel the helicopter floor vibration inputs in three orthogonal directions.

**Results and Discussions**

For demonstration purpose, a Bell-412 helicopter non-armored pilot seat frame was installed on the active seat mount system to verify the performance for pilot whole-body vibration reduction. The experimental setup consisted of three manikins of varying percentiles seated on the Bell-412 helicopter seat and was tested on a large mechanical shaker table. The shaker table provided representative floor vibration profiles of the Bell-412 helicopter, which included the first eight N/rev harmonic peaks of the main blade rotating frequency with the addition of a broadband random vibration profile. The active seat mount system was oriented at a 45 degree angle to the actuation direction of the single-axis shaker table to represent simultaneous vibration excitations in both X and Y directions. The tests showed that the multi-axis active seat mount system was able to reduce the vibration experienced by the manikins simultaneously at the head and body locations in both X and Y directions, with the 1/rev and 2/rev peaks reduced by more than 95%. This performance was also found consistent across all three manikins used in these tests.

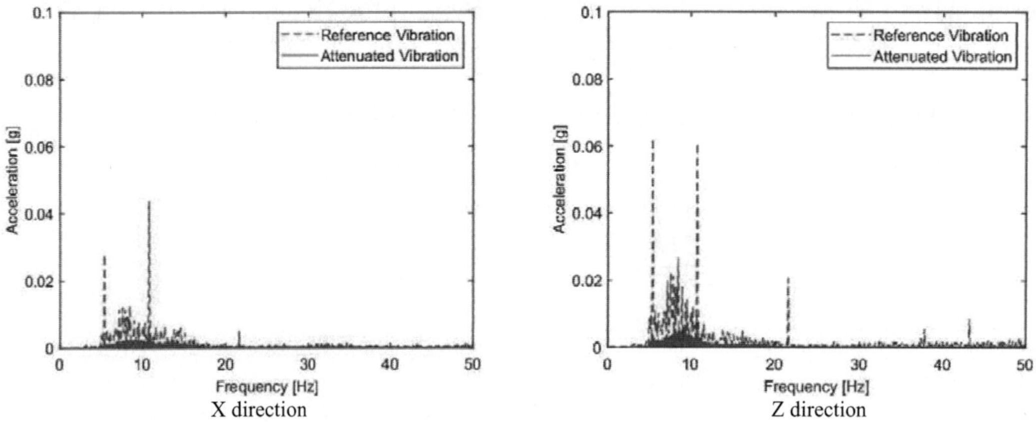


**Figure 1**. Vibration levels at the seat cushion interface per ISO2631-1

These results demonstrate that the multi-axis seat mount system is not only capable of reducing vibrations experienced by the pilot in single directions which has been shown in previous tests, but also capable of reducing vibrations experienced by the pilot in multiple directions. Therefore, the multi-axis active seat mount system can suppress the whole-body vibration exposure of the helicopter aircrew to improve their ride experience and enhance the protection on their health.

## References


*Chen Y, Wickramasinghe V and Zimcik DG (2011) Development of adaptive helicopter seat for aircrew vibration reduction. Journal of Intelligent Material Systems and Structures 22(5): 489-502.*

*Hadj-Moussa F, Fereidooni A, Chen Y et al. (2018) Development of a multi-axis active seat mount to mitigate vibration transmission to helicopter aircrew. In: 2018 AIAA/AHS Adaptive Structures Conference, Kissimmee, USA, 8-12 January 2018.*

# Experimental Characterization of Pull-in Parameters for an Electrostatically Actuated Cantilever

**Andrea Sorrentino*, Giovanni Bianchi, Davide Castagnetti, and Enrico Radi**

*Department of Science and Methods for Engineering, University of Modena and Reggio Emilia, Via Amendola 2, 42122, Reggio Emilia, Italy*

**andrea.sorrentino@unimore.it*

**Abstract**

Micro-electromechanical systems (MEMS) are a promising research frontier thanks to their multiple physical fields properties. In the field of microcantilever actuators, Radi et al., 2017, proposed an accurate analytical approach for estimating the pull-in characteristics of microcantilever actuators subject to electrostatic actuation. The present work assesses this previous analytical model via experimental tests with the use of a simple millimeter-scale device. The aim of the work is to measure the critical pull-in voltage and the deflection of an actuated cantilever beam for different configurations in order to validate the variation of the pull-in voltage with the geometrical parameters of the device provided by theoretical investigations. Preliminary tests show that the experimental pull-in voltage and deflection are in good agreement with the results provided by the analytical model. Specifically, the relative difference between experimental and analytical values of pull-in voltage is in the range between 0.7% and 10%.

## Introduction

This work experimentally investigates the response of an electrostatically actuated cantilever beam, which reproduces the typical behavior of the micromechanical switching blocks in MEMS applications. The interesting properties of the MEMS devices typically arise from the behavior of the active parts, which, in most cases, are in the forms of cantilevers (Ke and Espinosa, 2005). In particular, the importance of pull-in induced failures in these applications has received great attention in literature in recent years. At a critical voltage, referred to as 'pull-in voltage', the cantilever beam suspended tip pulls-in onto the above conductive substrate (Zhang et al., 2014). This work aims to experimentally assess the applicability of an analytical model from the literature (Radi et al., 2017), for the prediction of the pull-in voltage and the deflection of an electrostatically actuated cantilever beam. Specifically, we designed and built a millimeter-scale cantilever, which was actuated through an ad-hoc electric circuit. The experimental results closely match the predictions from the analytical model: in fact, the analytical deflection differs from the experimental measurement in the order of tens of microns.

## Methods

First the work focused on the design and prototype development of a millimeter-scale model of the MEMS device, where the intermolecular surface forces are negligible. The system is composed of two plates, the suspended electrode and the ground, both made of steel C100S. Both plates have a thickness, t=0.2 mm, a width, w=12.7 mm, and the total free length, l, of the suspended electrode was equal to 50 mm. The gap between the two electrodes was set equal to 0.6 mm, and obtained through a simple bi-adhesive tape (Fig. 1-a). The macro-scale dimensions of the device require a high actuating voltage to reach pull-in. Thus, we used a high voltage DC-DC power supply (EMCO CB101, output voltage from 0 to +10kV) managed through a manual potentiometer to actuate the device up to the pull-in threshold. In order to measure the tip deflection of the suspended electrode, we used a single point laser-doppler vibrometer (Polytec OFV-505 sensor head) linked to a National Instrument data acquisition board (NI 9211). An algorithm implemented in the LabVIEW environment registered and processed both the actuating voltage and the deflection measurements. Fig. 1-b shows the set-up of the test bench. Finally, we tested some different configurations of the cantilever to examine the influence of some parameters on the pull-in: specifically, we investigated a variable free length, l, from 50 to 70 mm, in combination with a gap from the ground, g, from 0.6 to 0.8 mm (see Fig. 1-b). For each configuration, we performed 10 replications of the tests.

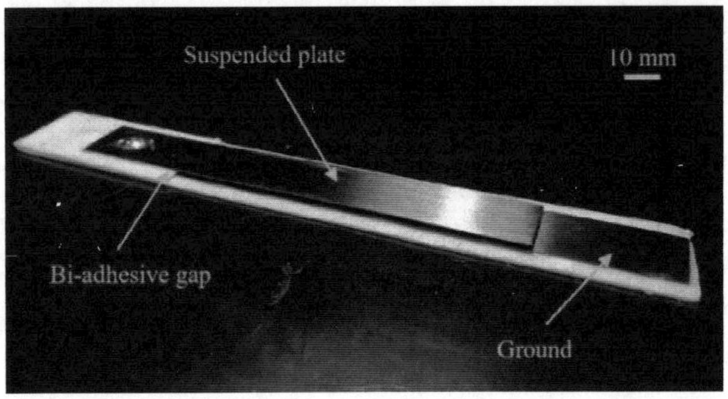


(a)

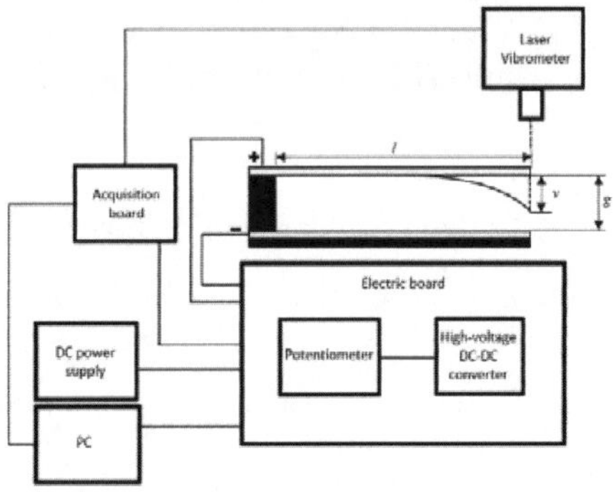


(b)

**Figure 1**. Millimeter scale device implemented (a), schematic of the testing benchmark (b)

## Results and Discussions

Table 1 compares the critical pull-in parameters for the different configuration of the specimens investigated where, $V_{PI}^{E}$ and $V_{PI}^{A}$, represent the experimental and the analytical pull-in voltages, respectively, and $v_{PI}^{E}$ and $v_{PI}^{A}$ the corresponding pull-in deflections.

**Table 1**. Comparison between the experimental and analytical pull-in voltage and tip deflection

| | **l = 50 mm** | | | | **l = 70 mm** | | | |
|---|---|---|---|---|---|---|---|---|
| **g [mm]** | $\mathbf{V_{PI}^{E}}$ [V] | $\mathbf{V_{PI}^{A}}$ [V] | $v_{PI}^{E}$ [mm] | $v_{PI}^{A}$ [mm] | $\mathbf{V_{PI}^{E}}$ [V] | $\mathbf{V_{PI}^{A}}$ [V] | $v_{PI}^{E}$ [mm] | $v_{PI}^{A}$ [mm] |
| 0.6 | 1250 ± 14.98 | 1342 | 0.252 ± 0.015 | 0.267 | 690 ± 2.45 | 685 | 0.280 ± 0.013 | 0.267 |
| 0.8 | 2040 ± 42.45 | 2066 | 0.290 ± 0.055 | 0.356 | 950 ± 8.35 | 1052 | 0.420 ± 0.023 | 0.356 |

The experimental results exhibit very good agreement with the analytical predictions from the current model. Additional theoretical methods able to predict the pull-in response of a micro-cantilever actuator are discussed in the review paper from Radi et al., 2017. In particular, we obtained a relative difference between experimental and analytical values of the pull-in voltage in the range between 0.7% - 10%, whereas the relative difference of the pull-in deflection falls in the range of 1.1% up to 18%. We can observe that the air gap and the free length have a great influence on pull-in instability. In particular, by increasing the gap, g, of 0.01 mm the pull-in voltage increases by about 18-36 V, depending on the free length.

## References


*Ke C and Espinosa HD (2005) Nanoelectromechanical Systems and Modeling. Handbook of Theoretical and Computational Nanotechnology.*

*Radi E, Bianchi G and di Ruvo L (2017) Upper and lower bounds for the pull-in parameters of a micro- or nanocantilever on a flexible support. International Journal of Non-Linear Mechanics 92 (July 2016). Elsevier Ltd: 176–186. DOI: 10.1016/j.ijnonlinmec.2017.03.011.*

*Zhang WM, Yan H, Peng ZK, et al. (2014) Electrostatic pull-in instability in MEMS/NEMS: A review. Sensors and Actuators, A: Physical 214. Elsevier B.V.: 187–218. DOI: 10.1016/j.sna.2014.04.025.*

# Design, Prototyping and Validation of a New PVDF Acoustic Sensor

**Andrea Sorrentino*, Yuri Ricci, Davide Castagnetti, and Luca Larcher**

*Department of Science and Methods for Engineering, University of Modena and Reggio Emilia, Via Amendola 2, 42122, Reggio Emilia, Italy*

*andrea.sorrentino@unimore.it*

**Abstract**

This work focuses on the design, prototyping and testing of a new millimeter-size 3D printed acoustic sensor based on a Polyvinylidene Fluoride (PVDF) film. The piezoelectric PVDF has been widely used for sensors development in many applications due their high sensitivity and low cost. The aim of this work is to design an acoustic sensor that collects the acoustic wave and convey the most part of it as a force on the piezoelectric PVDF film in order to increase the sensitivity of the microphone. The work presents a detailed analytical model describing the mechanical behavior of the system. Through a metaheuristic optimization algorithm, we found the optimal geometric parameters of the system that maximize the acoustic force on the piezoelectric film. Analytical results show that the proposed solution exhibits a good value of sensitivity in the frequency range 10-10000 Hz. The proposed acoustic sensor was manufactured through 3D printing in ABS material and the tests focused on investigating the sensitivity of the system at different frequencies.

**Introduction**

Sensor based on PVDF film are attractive thanks to their low cost and high sensitivity. When subject to a strain field, PVDF can generate voltages 10-20 times higher than piezoceramics, for the same pressure input. Microphone based on PVDF micro-pillar between patterned electrodes were proposed by (Xu et al., 2009). This sensor configuration presents high sensitivity values in combination with reduced capacitance, but with an expensive manufacturing cost. This work focuses on the design, development and optimization of a low cost PVDF acoustic sensor, which converts the acoustic wave pressure into a force on a small area of the PVDF film. The aim of the work is to maximize the output voltage on the opposite surfaces of the film while decreasing the device capacitance, that corresponds to an increase of sensitivity of the microphone. The first step of the work developed an analytical model of the mechanical response of the microphone. In the second step, through a constrained optimization algorithm, we calculated the geometrical dimensions of the system which maximize the force applied on the PVDF film due to the acoustic force. The third step involved the prototype development of the system, manufactured through FDM 3D printing technology with ABS plastic.

**Methods/Model Formulations**

Figure 1(a) shows the architecture of the proposed system. The acoustic wave acts on a circular plate supported by a cylindrical pillar, which conveys the acoustic pressure to the PVDF film. A cylindrical outer frame supports the mobile element through three rectangular cross-section flexural springs, which allows the mobile element to convey the acoustic load on the polymeric film, while maintaining the necessary stability. A preload system ensures the contact between the pillar and the film (Fig.1(a)), whichever the loading condition on the mobile equipment. In the first step, we developed an analytical model of the system, which includes both the mechanical and electrical model of the sensor and of the PCB circuit. The analytical model describes the microphone as a mass-springs system, considering all the force components acting on the device. In the second step, we addressed the optimization problem of maximizing the force applied on the PVDF film, named $F_{pz}$, due to the acoustic force, named $F_{ac}$, by acting on the following design variables: the width, $w$, the thickness, $t$, and the length $l$ of the flexural springs, for a maximum acoustic pressure, $p_{max}$, equal to 300 Pa (Fig.1(a)). The optimization involves six constraints, among which the resonance frequency, the yield strength and the buckling load, and used a penalty-constrained Simulated Annealing (SA) algorithm implemented in Matlab. In the last step, we manufactured a prototype of the sensor using 3D printing with ABS material (Fig.1(b)).

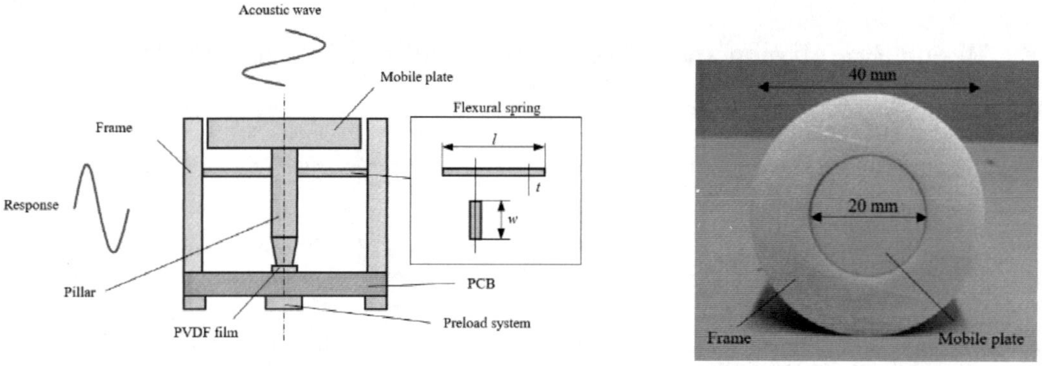


(a) (b)

**Figure 1**. Schematic of the PVDF acoustic sensor (a), top view of the sensor's prototype make of ABS (b)

## Results and Discussions

Table 1 reports the optimal values of the design variables with the corresponding values of the objective function, $F_{pz}$, and the theoretical sensitivity of the sensor, $S^*$, provided by the optimization procedure for a maximum acoustic pressure, $p_{max}$, equal to 300 Pa.

**Table 1**. Optimization results

| | |
|---|---|
| $w$ (m) | 0.001 |
| $t$ (m) | 0.005 |
| $l$ (m) | 0.010 |
| $F_{ac}$ (N) | 0.094 |
| $F_{pz}$ (N) | 0.048 |
| $S^*$ (μV/Pa) | 700 |

The results show that only a fraction of the acoustic force is transmitted to the polymeric film. The value of the sensitivity, $S^*$, is affected by the resonance frequency constraints: the imposed frequency range strongly affects the optimization search and consequently the dimensions of the design variables of the springs. Compared to the simple flat configuration proposed in (Xu et al., 2009), with a theoretical sensitivity equal to 3.5 μV/Pa, this solution features a remarkably strong improvement, with a sensitivity two order of magnitude higher, i.e. 700 μV/Pa. Preliminary experimental tests to validate this model have been carried out. The work shows how to design, develop and optimize a low cost PVDF acoustic sensor.

## References

*Xu J, Dapino MJ, Gallego-Perez D, et al. (2009) Microphone based on Polyvinylidene Fluoride (PVDF) micro-pillars and patterned electrodes. Sensors and Actuators, A: Physical 153(1): 24–32. DOI: 10.1016/j.sna.2009.04.008.*

**Gap in pagination due to unavailable papers.**

**Pages 73-74**

# Design and Experimental Validation of a Haptic Glove and Robotic Arm Powered by Magnetorheological Actuators for Dexterous Object Manipulation in VR

**Benjamin Bedard***, **Louis-Philippe Lebel, Bruno-Pier Busque, Marc Denninger, Guifré Julio, Alexandre Girard, and Jean-Sébastien Plante**

*Faculté de génie, Université de Sherbrooke, 2500 Boulevard de l'Université, Sherbrooke, QC, Canada*

**Benjamin.bedard@usherbrooke.ca*

## Introduction

Virtual Reality (VR) is increasingly used in a multitude of computer applications such as gaming, training for dexterous manual tasks surgical interventions, and simulations of complex group missions in military or space exploration [1]. Even though VR creates amazing visual experiences, it still misses the sense of touch (haptic) [2]. With the addition of a high-fidelity haptic system for the hand, VR has the potential to become a total immersion experience. At term, users could touch and manipulate objects like in the real world. However, today's haptic VR systems are limited by their actuation technologies, and cannot provide a high-quality force feedback to both hands while being lightweight, robust and cost effective [3].

A full haptic VR system consists in the combination of a haptic exoskeleton glove that gives grip sensation and a collaborative robotic arm that generates forces in operational coordinates to simulate gravity, contact forces, and object inertial reactions. Existing haptic VR system like CyberForce have low fingertip forces and low Cartesian forces compared to human capabilities [4]. Force-feedback gloves (FFG) such allows users to pick up and feel a virtual object in a natural way by providing fingertips forces. However, when using conventional gear-motor actuators, such FFG face an inevitable trade-off between quality of force and force density (device size). Devices end-up being either lightweight but with limited dynamic responses due to the inertia of the actuator system [5] and/or having low fingertip forces, for example, of only 5 N for the Dexmo device [6]. Robotic arms are also designed with gear-motors and thus, face the same performance trade-off. Today's best robotic arms may present sufficient force capabilities to interact with a human arm, but they lack the dynamic response [5]. Laboratory demonstrations showed that Tendon-Driven Manipulator actuated by Magnetorheological (MR) clutches (TDM-MR) have excellent dynamic response given by the clutches' low output inertia combined with their high force density [5]. Also, the approach is simple and cost effective since it does not rely on high-precision machined parts (e.g. harmonic drives). This paper investigates the potential of the TDM-MR approach for VR interaction by designing and evaluating the performance of a 4-DOF force-feedback glove coupled to a 3-DOF robotic arm.

## System Design

An ideal haptic glove and robotic arm system for VR applications should have performance levels equal or better than humans. Anthropomorphic design requirements for VR applications are summarized in Table 1.

The proposed prototype system consists of a 4-DOF haptic exoskeleton glove combined with a 3-DOF robotic arm (see Figure 1). The 4-DOF glove is actuated by an array of 4 miniatures MR clutches pulling 4 fingers: the thumb, the index, the middle and the ring. The miniature MR clutches have a unique design feature consisting of using lightweight non-magnetic shear surfaces, which provides the advantages of: (1) minimizing the inertia of the clutch output rotor and (2) easing machining and assembly. The 3-DOF robotic arm uses a single industrial motor distributing power to 3 active DOF each using a pair of antagonistic MR clutches actuators for a total of 6 MR clutches.

## Experimental Results

Most of the requirements from Table 1 have been measured and suggest the TDM-MR approach to be effective. Fingertip forces of 32N are measured and minimum force of 50 N at the end-effector is confirmed experimentally. Dynamic response of 33 Hz has been measured for the arm and 60 Hz for MR clutches. Thus, blocked-force bandwidth of the glove's fingers are expected to be above 10 Hz. In terms of reliability and robustness, the robotic arm has been used intensively more than 10 hours of operation. Thus, the TDM-MR as the potential to be a reliable technology for general collaborative robotics applications.

**Table 1**. Ideal design requirements for a haptic glove/arm system

| Spec. | Description | Value | Units | Flexibility | Ref. |
|---|---|---|---|---|---|
| A | Force generated at fingertip (glove) | 30 | N | Minimum | [7] |
| B | Cartesian force generated at end-effector at any time (robotic arm) | 50 | N | Minimum | [8] |
| C | Dynamic response (glove & robotic arm) | 10 | Hz | Minimum | [9] |
| D | Reliability (glove & robotic arm) | | | | |
| E | Resist to impacts (glove & robotic arm) | | | | |

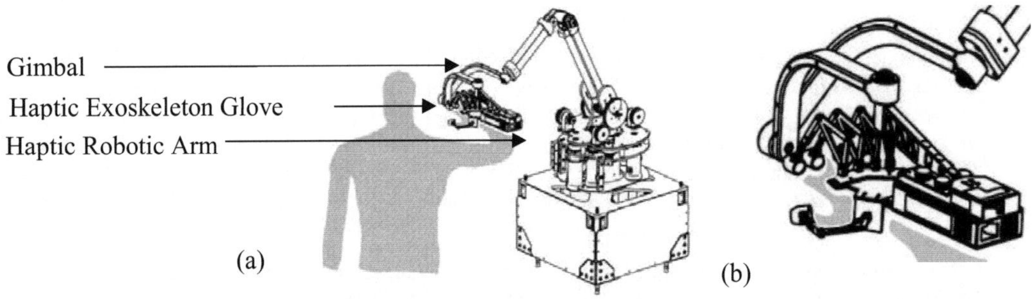


**Figure 1.** (a) Haptic VR system; (b) haptic exoskeleton glove

## Conclusion

This paper presents a feasibility study of TDM-MR for VR applications by evaluating the performance of a combination of a 4-DOF haptic exoskeleton glove and a 3-DOF haptic robotic arm. Performance of the haptic robotic arm has been successfully tested and shown to meet requirements of force levels, dynamic response, robustness, and impact resistance. Thus, the haptic glove is expected to meet ideal design requirements. Already, the results suggest the TDM-MR approach to be a relevant technology for collaborative robotic applications with the highest performance requirements such as VR touch feedback. The result could generalize to all collaborative robotic applications, making TDM-MR a high performance robotic solution for man-machine interaction.

## References

*[1] « Virtual Reality in Entertainment », https://thinkmobiles.com/blog/virtual-reality-entertainment/.*

*[2] J. Blake et H. B. Gurocak, « Haptic Glove With MR Brakes for Virtual Reality », IEEEASME Trans. Mechatron., vol. 14, no 5, p. 606‑615, oct. 2009.*

*[3] M. Bégin et al., « Experimental Assessment of a Controlled Slippage Magnetorheological Actuator for Active Seat Suspensions », IEEEASME Trans. Mechatron., vol. 23, no 4, p. 1800‑1810, août 2018.*

*[4] « CyberForce »,Cyber Glove System LLC. http://www.cyberglovesystems.com/cyberforce.*

*[5] J. Viau, P. Chouinard, J. L. Bigué, G. Julió, F. Michaud, et J. Plante, « Tendon-Driven Manipulator Actuated by Magnetorheological Clutches Exhibiting Both High-Power and Soft Motion Capabilities », IEEEASME Trans. Mechatron., vol. 22, no 1, p. 561‑571, févr. 2017.*

*[6] « Dexta Robotics - Touch the Untouchable ». Disponible sur: https://www.dextarobotics.com/en-us.*

*[7] N. Miyata, K. Yamaguchi, et Y. Maeda, « Measuring and modeling active maximum fingertip forces of a human index finger », in 2007 IEEE/RSJ 2007, p. 2156‑2161.*

*[8] « Armeo®Power », Hocoma. .*

*[9] S. R. Ellis, « What are virtual environments? », IEEE Comput. Graph. no 1, p. 17‑22, janv. 1994.*

# Experimental Assessment of a Linear Actuator Driven by Magnetorheological Clutches for Automotive Active Suspensions

**William East***, **Jérôme Turcotte, and Jean-Sébastien Plante**

*Créatek, Université de Sherbrooke, 3000 boul. de l'Université, Sherbrooke, QC, Canada, J1K 0A5*

**William.East@USherbrooke.ca*

## Introduction

Recent advances in self-driving vehicles are reshaping the future of personal transportation. Manufacturers, such as Volvo, are presenting autonomous concept cars aiming to change intercity travel by replacing heavily polluting short flights by self-driving living spaces where people can work and sleep (Volvo, 2019). A fundamental challenge to such advanced travel concepts is motion sickness due to road vibrations. Studies show that when passengers in moving vehicles are not focused on the road while doing an active intellectual activity such as drawing or reading, they become much more subject to vibration/motion induced sickness (Diels and Bos, 2016). An effective way to solve this problem is with semi-active or active suspensions that can cancel road vibrations and correct vehicle attitude during maneuvers. Although mechanically simpler, semi-active suspensions are fundamentally limited because they can't provide any vehicle attitude corrections and have limited capabilities for ride quality control with high frequency harshness due to the limited switching rate of the control system (Rajamani, 2006). Hence, there is an increasing interest amongst automotive manufacturers for fully active suspension technologies.

For the last 30 years, researchers have attempted to develop fast, strong and reliable active suspension systems for on-road cars to increase safety, comfort and vehicle performance, and all that at low costs. Table 1 presents either published quantitative values, or inferred qualitative statements evaluating the main active suspension technologies developed so far with respect to functional requirements. All reported systems have at least one problematic requirement based on intrinsic trade-offs which explains why these technologies are not widely available on the market today. Hydraulic systems are powerful, but relatively slow and expensive. Electromagnetic direct-drive systems have limited forces and/or are heavy and expensive while electromagnetic hydrostatic and electromagnetic geared systems both have bandwidth limitations due to gearing.

**Table 1**. Functional requirements of active suspension technologies

| Functional Requirement | Hydraulic | Electromagnetic Direct-drive | Electromagnetic Hydrostatic | Electromagnetic Geared |
|---|---|---|---|---|
| Companies | Toyota, Nissan, Mercedes, GM | Bose, Eindhoven University | ClearMotion | Audi |
| Max. Force (N) | High | 1000 | N/D | 3700 |
| Max. Linear speed (m/s) | Low | 1 | 1.5 | N/D |
| Bandwidth (Hz) | 5 | >100 | 10 | 5 |
| Power consumption (W) | 3000-5000 | High | 500 | 60-250 |
| Mass and inertia (kg) | N/D | N/D | Low | N/D |
| Cost ($) | High | High | N/D | N/D |
| Vehicle integration | Good | Good | Good | Good |
| Reliability | Low | N/D | N/D | High |
| References | (Venhovens, 1993) (Gysen et al., 2009) (Martien, 2018) | (Jones, 2005) (Gysen et al., 2009) (T.P.J., 2011) | (ClearMotion, 2016) | (Audi, 2017) (Adcock, 2017) |

*Green cell = Requirement is met, Red cell = Not met, Grey cell = No information

Studies on an alternative actuation technology based on continuously slipping clutch actuators using magneto-rheological (MR) fluids show that this approach has a great potential for active vibration control (Bégin, 2018) & (Chouinard, 2014). MR clutches work by engaging an input surface to an output surface separated by a layer of magnetorheological fluid (MRF). The MRF viscosity, and thus, the transferred torque, is controlled by varying a magnetic field passing through the fluid. Doing so, the layer of fluid of the MR clutch actuator decouples the input inertia (motor and gearbox) from the system's output (wheel), thereby making actuator inertia so low that it can be considered as a pure force source.

This paper presents a feasibility study of an active suspension strut powered by MR clutches. A prototype is designed and built (Figure 1) to replace the front suspension struts of a 2001 BMW 330Ci. Functional requirements for force, linear speed, bandwidth and intrinsic parasitic forces are established from a vehicle dynamic simulation within the Siemens LMS Amesim Imagine Lab software and from on-car accelerometer measurements. Figure 2 shows the actuator's working principle. A design concept of an active strut is presented using a 6.5kW high-speed brushless motor to feed a pair of counter rotating MR clutches providing upward and downward forces on a gear-rack mechanism. A full-scale prototype is fabricated and extensively characterized in a laboratory. First, a rotary test bench is used to characterize the MR clutches and high-speed electric motor independently of the gear rack mechanism. Then, an experimental characterization of the full system is conducted on an MTS dynamometer to assess the force, linear speed, bandwidth, and intrinsic parasitic forces of the actuator prototype.

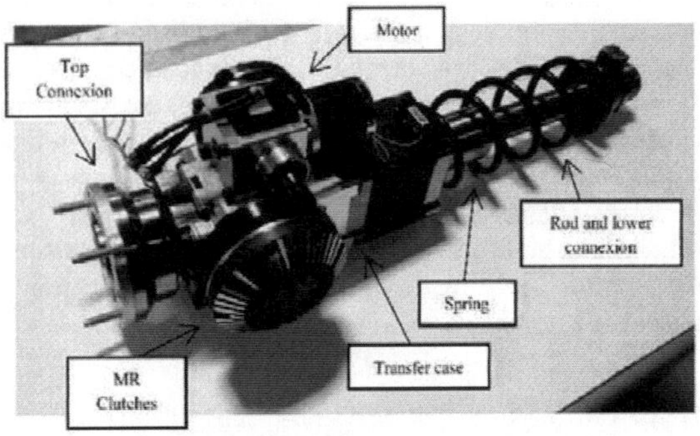


Figure 1. Actuator prototype

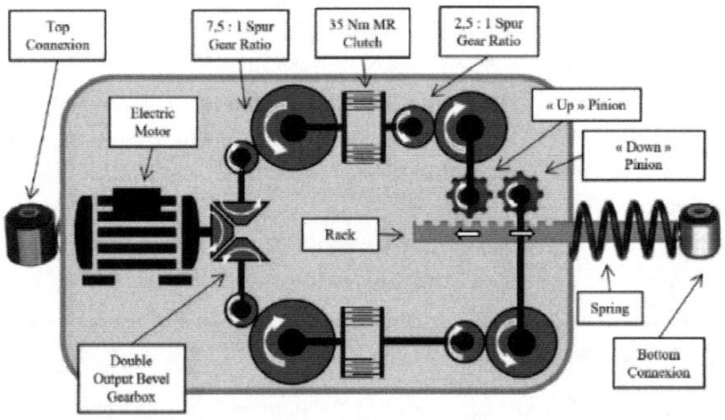


**Figure 2**. Working principle diagram of the MR clutch actuator

Results from the motor test bench show that the selected KDE700XF-295 electric motor outputs a maximum continuous power of 6.5 kW and a continuous torque of 3.8Nm at 16 500 RPM. The MR clutch outputs 36 Nm at 14 A, with a 29 Hz bandwidth and parasitic torque of 2.5 % of the maximum output torque at a 2500 RPM slip rate. In the actuator reference frame, these results suggest a maximal force of 4800 N at a speed of 1.73 m/s, a bandwidth of 29 Hz and maximum parasitic forces from the MR clutch of 120 N. MTS dynamometer tests will be performed in June to confirm those values, and results will be added in the final paper.

**References**

*Adcock, I., 2017. Audi details new A8 active suspension [Website]. Sae International. URL*

*Audi, 2017. Audi A8 - Active Chassis [Website].*

*Bégin, M.-A., 2018. Experimental Assessement of a Controlled Slippage MR Actuator for Active Seat.IEEE/ASME.*

*Chouinard, P., 2014. Conception et validation expérimentale d'un système d'embrayages MR [...].*

*ClearMotion, 2016. ClearMotion [automobile ride control system], in: 2016 IEEE Hot Chips 28. Symposium (HCS).*

*Diels, C., Bos, J.E., 2016. Self-driving carsickness. Applied Ergonomics 53.*

*Gysen, B.L.J., Janssen, J.L.G., 2009. Design aspects of an active suspension system for automotive. IEEE Trans.*

*Jones, W.D., 2005. Easy ride: Bose Corp. uses speaker technology to give cars adaptive suspension. IEEE.*

*Martien, J., 2018. History of Active Suspension [Website]. DSC Sport.*

*Rajamani, R., 2006. Vehicle Dynamics and Control, Springer.*

*T.P.J., V. der S., 2011. Control of an automotive electromagnetic suspension system. Eindhoven.*

*Venhovens, P.J., 1993. Optimal control of vehicle suspensions. Technical University Delft.*

*Volvo, 2019. Volvo concept 360c - a new way to travel [Website].*

30th International Conference on Adaptive Structures and Technologies

# Embedded SiC Fiber Sensor Based Low-velocity Impact Monitoring of Composite Structures

**Hyunseok Kwon, Yurim Park, and Chun-Gon Kim***

*Department of Aerospace Engineering, KAIST, 291 Daehak-ro, Yuseong-gu, Daejeon, 34141, Republic of Korea*

*cgkim@kaist.edu

**Abstract**

Piezoresistive fibers can be a sensing element for structural monitoring of composite structures while minimizing the deterioration of mechanical properties of host structures. Among piezoresistive fibers, the semi-conductive SiC fibers have excellent strain sensitivity due to inherent characteristics of semi-conductive materials. In this study, Embedded semi-conductive SiC fiber sensor based low-velocity impact monitoring of composite structures was proposed. The SiC fiber sensor network was embedded into composite panel during its fabrication process. The resistance change of the embedded SiC fiber sensor was measured by converting the resistance change to voltage change through a Wheatstone bridge. By using an Impact hammer, impacts were introduced to the SiC fiber sensor embedded composite panel, and impact signals were acquired using embedded SiC fiber sensor. The impact signals were successfully acquired. The acquired impact signals showed high repeatability at a same impact points and high location distinguishability for different impact points.

**Introduction**

Embedding of sensors during the fabrication process can be easily achieved in the case of composite materials. The sensors that are embedded into a composite material are not only more sensitive to inside damage, but also protected from external environments, enabling long-term structural monitoring. However, existing sensors have different mechanical properties and geometry from components of composite materials, resulting in degradation of mechanical properties of the structure. Therefore, there have been efforts to develop a new type of sensor to overcome this problem. Of the new sensor types, sensing mechanisms that utilize the piezoresistive properties of conductive fibers such as carbon fibers are of interest lately. Conductive fibers, such as carbon fibers, have the same geometry as the reinforcing fibers that make up the composite material, thus minimizing the influences on the mechanical properties, as well as supporting the load at the same time. Therefore, the sensing system using the piezoresistive properties of the fiber can be a sensor type suitable for effective and consistent monitoring of composite structures. In previous research, a study to utilize semi-conductive SiC fibers as a sensor was performed [1]. In this study, we propose an embedded SiC fiber sensor network based low-velocity impact monitoring for composite structures.

**Methods**

In this study, semi-conductive SiC fibers fabricated by Ube Industries, Ltd., Japan were used as a sensor. A yarn of the SiC fibers was used as a sensing element. Composite panels were fabricated using glass/epoxy prepregs (GEP 118, Muhan Composite Co., Ltd.), and SiC fiber sensors were embedded between prepreg plies during the fabrication process. A total of four SiC fiber sensors were embedded into the composite panel. Fig. 1 (a) shows the geometry of composite panel and placement of the embedded SiC fiber sensor. For the connection of the embedded SiC fiber to the external electric wire, the end of the embedded SiC fibers was filled with conductive resin (silver paste). Fig. 1 (b) shows the cured specimen. The resistance change of the embedded SiC fiber sensors was measured by converting the resistance change to voltage change using the electrical circuit based on a Wheatstone bridge. The SiC fiber sensor network embedded composite panel was clamped at four corners. Random impact were applied to the composite panel using an Impact hammer, and impact signals were acquired based on the resistance change of embedded SiC fiber sensors.

a) 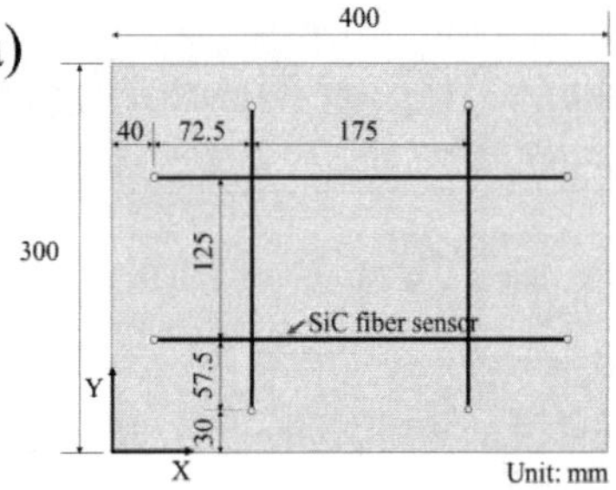 b) 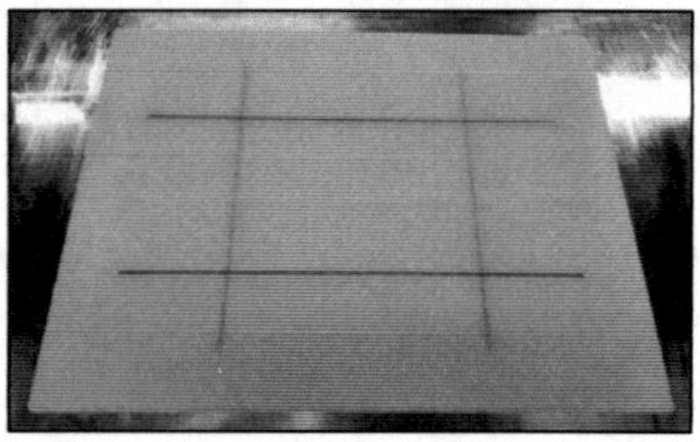



**Figure 1**. **a)** Geometry of SiC fiber sensor embedded composite panel and **b)** Cured specimen

**Result and Discussion**

Impacts were introduced at two different points, and impact signals were acquired. The impacts were applied at point A (150 mm, 150 mm) and point B (250 mm, 200 mm), and the impact was applied three times at each point. Impact signals were successfully acquired using the embedded SiC fiber sensors, and Fig. 2 shows the acquired impulse signal. Repeatability in impact signal acquisition at a same impact point and signal distinguishability according to impact locations were evaluated. In order to evaluate the repeatability and distinguishability between acquired impact signals, correlation coefficients between the signals were calculated. The correlation coefficient between two different impact signals ($S_1$ and $S_2$) was calculated using Eq. (1). The value of correlation coefficient is between -1 and 1. A value of correlation coefficient closer to 1 signifies a, higher similarity between the two signals and a value closer to 0 signifies lower similarity between the two signals. As a result of correlation coefficient calculation, correlation coefficient values between impact signals at a same point showed very high similarity of 0.99 or more. The impact signals acquired at different points showed low similarity with correlation coefficient values of 0.76 or less. Therefore, it was shown that the obtained signals have high repeatability at the same position and the signals obtained at different points have position discrimination.

$$Correlation\ coefficent\ value = \frac{\Sigma(S_1 - \bar{S_1})(S_2 - \bar{S_2})}{\sqrt{\Sigma(S_1 - \bar{S_1})^2 (S_2 - \bar{S_2})^2}} \tag{1}$$

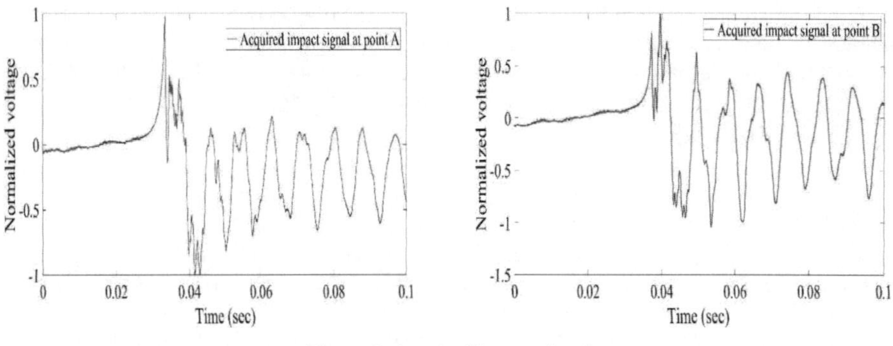



**Figure 2**. Acquired impact signals

**References**

*Kwon H, Park Y, Shrestha P, and Kim CG (2018) Application of silicon carbide fibers as a sensor for low-velocity impact detection and localization. Structural Health Monitoring Online published.*

**Gap in pagination due to unavailable papers.**

**Pages 81-84**

# Microcantilever Sensor for Biochemical Reaction Detection

**Shervin Foroughi*, Jeetender Amritsar, Durai Chelvan,**
**Shanmugamsundaram Packirisamy, and Muthukumaran Packirisamy**

*Department of Mechanical, Industrial and Aerospace Engineering, Concordia University, 1515 Sainte-Catherine St. W. Montreal, QC, Canada*

**s_foroug@encs.concordia.ca*

**Abstract**

This paper presents the investigation of a Polyvinyl-Di-Chlorofluoride (PVDF) aluminium coated micro-cantilever as a mechanical sensor. The reaction between the HSP90 protein and ATP is sensed by the cantilever due to the HSP90 conformational changes which deflects the cantilever. This demonstrates the effective usage of the micro-cantilever in investigation of biochemical reactions.

**Introduction**

Micro Electro Mechanical Systems (MEMS) can perform multiple functions such as actuation and sensing in a micro/molecular scale in order to either measure or control of the physical, chemical or biological parameters. The glucose monitoring in diabetes patients, body fluids screening for disease detection, drug discovery, biomarker sampling/analysis, protein-protein/ligand interaction, antigen-antibody binding are the examples for biomedical applications of MEMS molecular sensors (Zhang, 2014; Chandrasekaran, 2008). A class of MEMS sensors are the microcantilevers with immobilized surface which have been widely used for detection of chemical molecules, DNA or protein molecules by monitoring of the microcantilever bending displacement (Zhu, 2011). In this research the biochemical reaction between antibody-antigen with various concentration is investigated by optical monitoring of microcantiliver deflection.

**Methodology**

Hsp90 is a ubiquitous chaperone in the cell which assists in folding or unfolding of other proteins to maintain steady under imposed positive thermal stress, and to help for degradation. It can attach to the ATP and helps for ATP hydrolysis in the folding reaction (Isaacs, 2016) Figure 1 illustrates the schematic of this process.

In order to identify the biochemical mechanism of the HSP90 with involvement of the ATP as well as resulted conformational changes, the aluminum coated PVDF microcantilevers have been utilized. The experiments were performed with HSP90 concentrations of 0.095, 0.182 μg/μl. The ATP was utilized with the same concentration of 0.254 μg/μl in all tests. In the first step, sort of experiments were carried out by deposition of individual HSP90 and ATP droplets with volume of 0.2 μl, on the surface of cantilever. Next, the ATP, with the constant concentration of 0.254 μg/μl, and HSP90 with above mentioned various concentrations were mixed separately. Then, the sample droplets in 0.2 μl volume were prepared and deposited on the cantilevers. In all experiments the droplets were deposited on the surface of cantilevers near to their tip. The size of cantilevers was 6 mm in length, 2.5 mm in width, and 25 μm in thickness. The surface contact of the droplet and cantilever after evaporation was approximately between 1.5 to 2 mm. Due to the weight of the droplet, evaporation process and reaction between HSP90/ATP, a cantilever was experiencing the compression and tension bending during a certain period of time until reaching the equilibrium state. In this process, the deflection of cantilever was being recorded by employing the optical lever method, as illustrated in Figure 2.

**Results and Discussions**

The results of experiments are shown in Figure 3. Figure 3a displays no permanent effect on the deflection of the cantilever in case of ATP droplet deposition. In fact, after the evaporation of the ATP droplet, the cantilever came back to its initial position. However, the deformation of cantilevers after evaporation of deposited HSP90 droplets is noticeable. This effect happened due to the interaction of the HSP90 with the surface of the cantilever as well as its conformational transition, as shown in Figure 3b. Figure 3c illustrates the higher magnitude of deflections in presence of HSP90/ATP mixture. This happened because of the interaction between HSP90-ATP and the resulted conformational changes of the protein. The results depict by increasing the concentration of the HSP90, the deflection of the cantilever increases which indicates the effect of concentration on the reaction between ATP and HSP90. These results clearly reveal the effective usage of the micro cantilevers in biochemical studies.

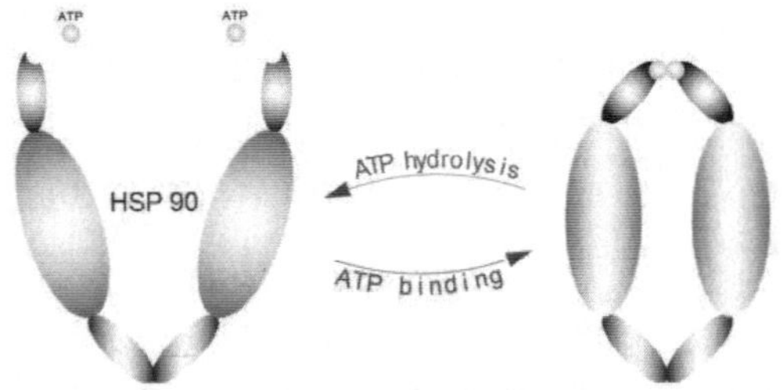


**Figure 1**. Schematic of the binding and hydrolysis of ATP by HSP90 and conformational changes

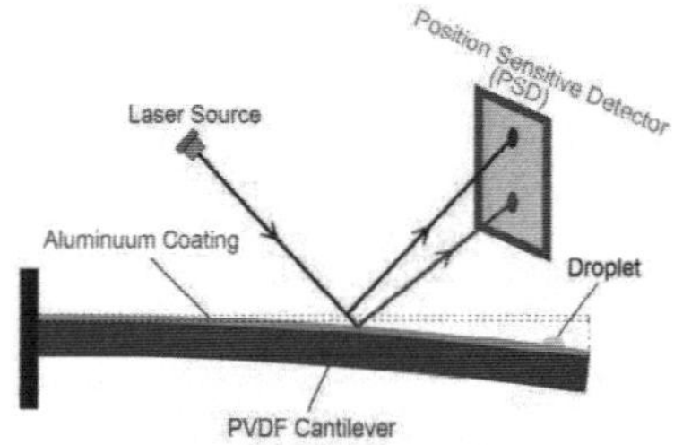


**Figure 2**. Schematic for configuration of cantilever and optical lever method

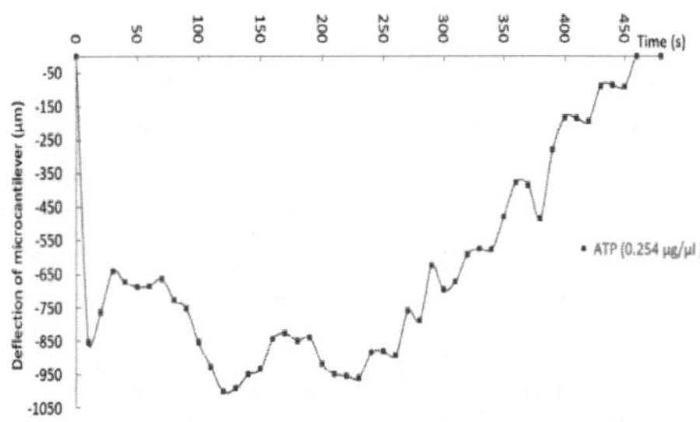


a) Deposition of ATP alone

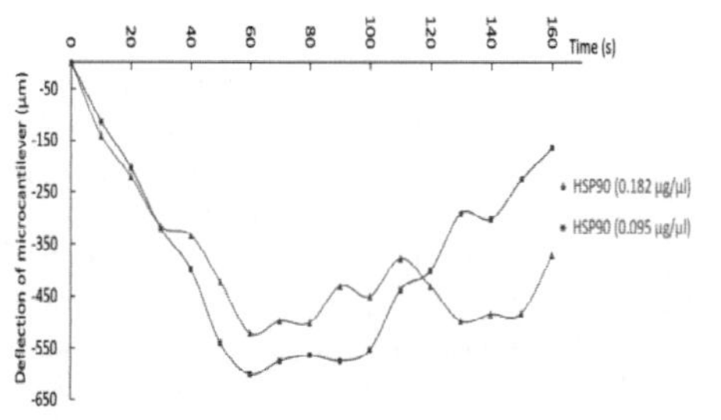


b) Deposition of HSP90 alone with two different concentrations

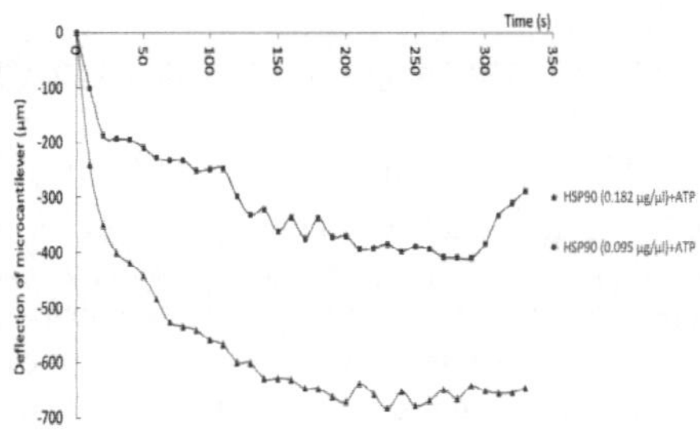


c) Deposition of HSP90/ATP mixture

**Figure 3**. Recorded deflection of cantilevers due to deposition of HSP90, ATP and HSP90/ATP droplets

## References

*Chandrasekaran A., Packirisamy S., You J. L., Acharya A., Packirisamy M., and Maxwell D. (2008) Bioresistive identification of heat shock protein 90. Biomicrofluidics 2(3): 034102.*

*Isaacs J. and Whitesell L. (2016) Hsp90 in cancer: beyond the usual suspects (Vol. 129). Academic Press.*

*Zhang J.X.J. and Hoshino K. (2014) Molecular sensors and nanodevices: principles, designs and applications in biomedical engineering, Elsevier, Massachusetts.*

*Zhu Q. (2011) Microcantilever sensors in biological and chemical detections. Sensors & transducers, 125(2): 1-21.*

30th International Conference on Adaptive Structures and Technologies

# Control of a Full-scale Portion of a Regional Aircraft Morphable Wing

**Duc-Hien Nguyen[1], Ruxandra Mihaela Botez[1*], Teodor Lucian Grigorie[1,2], Mahmoud Mamou[3], and Yousef Mébarki[3]**

[1]*École de Technologie Supérieure, Montréal, Québec H3C 1K3, Canada*

[2]*Military Technical Academy "Ferdinand I", Bucharest 050141, Romania*

[3]*National Research Council, Ottawa, Ontario K1A 0R6, Canada*

[*]*ruxandra.botez@etsmtl.ca*

**Abstract**

The paper deals with a new morphing wing application developed with a full-scale portion of a regional aircraft. Shown are: the design of the control system for the actuation mechanism based on four miniature high-force in-house developed electrical actuators, the testing of the controlled morphing wing on bench and wind tunnel, and the evaluation of the aerodynamic gain of the morphing wing technology on the experimental model (pressure signals analysis, and infra-red thermography visualizations).

## Introduction

The work is developed in a research project, realized at Ecole de Technologie Supérieure in Montréal, Canada, in collaboration with Thales, Bombardier Aerospace, École Polytechnique, National Research Council Canada (IAR-NRC), and, also, with Italian researchers from University of Naples Frederico II, CIRA and Alenia, where a wing-aileron prototype is designed, tested and validated using wind tunnel tests at National Research Council Canada (IAR-NRC). The project aims to reduce the operating costs for the new generation of aircraft through in-flight fuel economy, and to improve aircraft performances, expand its flight envelope, replace conventional control surfaces, reduce drag to improve range and reduce vibrations and flutter risk. An objective for the research team was to develop a system for active control of the morphing wing during flight to move the transition point from laminar to turbulent flow closer to the trailing edge, reducing drag over an operating range of flow conditions univocally defined by mean of Mach numbers, angles of attack, and aileron deflection angles. The wing used in this project is a full-scaled portion of a real wing (Fig. 1a), having maximum chord of 1.5 m and a minimum one of 1.08 m. It consists of wing spars, ribs and an aileron (Fig. 1b). The controlled actuation system includes four similar miniature electromechanical actuators, disposed in two parallel actuation lines.

(a) (b)

**Figure 1**. Used wing as a full-scale portion of a regional aircraft wing.

## Methods/Model Formulations

The upper surface of the wing is a flexible one, being closed to the wing tip; the flexible skin is made of composite materials. Each of the four actuators modifies the flexible upper surface of the wing through direct actuation so that the upper flow is changed and consequently the transition point from laminar to turbulent flow is delayed. The actuators are fixed on the wing ribs, and at the top are attached to the flexible skin with screws; they would push or pull the flexible skin using the necessary torque until the desired vertical displacement of each actuator is achieved. The four vertical displacements of the actuators, correlated with the new shape of the wing, are provided by a database obtained through a preliminary aerodynamic optimization for different flight conditions. The control system is designed to control the positions of the actuators in real time, in order to obtain and to maintain the desired shape of the wing for a specified flight condition (Fig. 2). The feasibility and the effectiveness of the developed control system are demonstrated experimentally through the wind tunnel testing of the morphing wing model. On the extrados 32 Kulite pressure sensors are installed to sense the static pressure on the wing. Simultaneously with the control system characteristics monitoring, the user visualized on a parallel screen the real time Fast Fourier Transforms (FFT) associated to the 32 Kulite pressure sensors equipping the upper surface flexible skin (Fig. 3). As a secondary method to evaluate the transition point position over the entire wing model surface for each tested flow case the infra-red (IR) thermography was used (Fig. 4).

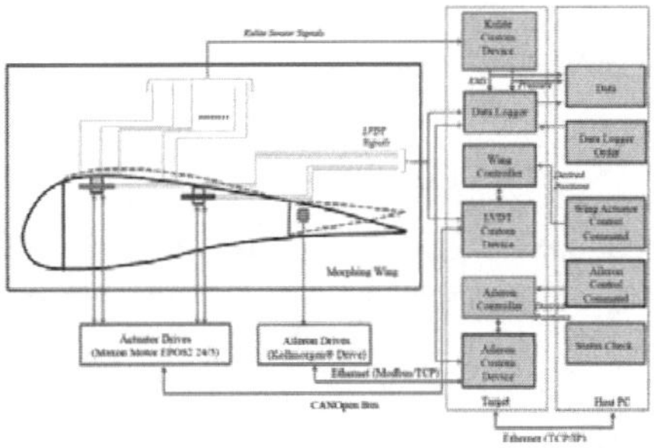

**Figure 2**. General architecture of the morphing wing control system.

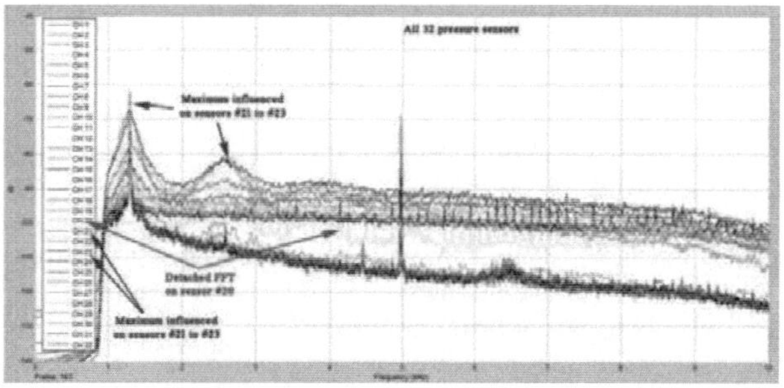

**Figure 3**. FFT of the pressure data acquired for Mach=0.15, α=-2°, δ=-2° flow case in morphed configuration.

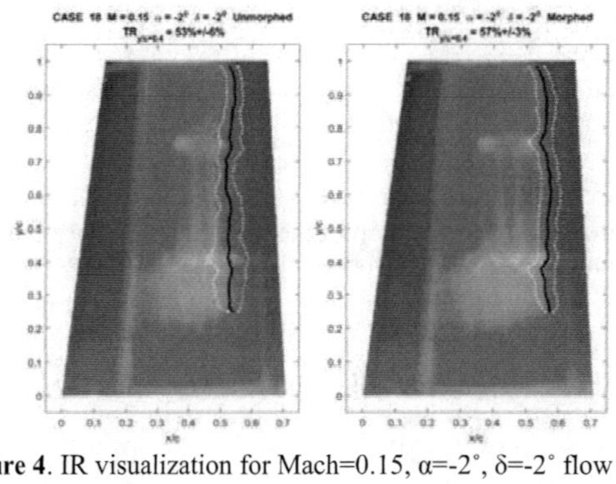

**Figure 4**. IR visualization for Mach=0.15, α=-2°, δ=-2° flow case.

## Results and Discussions

The results obtained during the wind tunnel testing proved that the morphing technology is one of the most promising technologies for the improvement of the aircraft aerodynamics, providing in this way a fuel consumption reduction for the next generation of aircraft. The research team observed that the morphing technology improved the average position of the laminar to turbulent flow transition over the whole wing with more than 2.5% of the wing chord for the great majority of the 97 tested flight cases.

## References


*Tchatchueng Kammegne MJ, Botez RM, Mamou M, Mebarki Y, Koreanschi A, Sugar Gabor O and Grigorie TL (2016) Experimental wind tunnel testing of a new multidisciplinary morphing wing model. 18th International Conference on Mathematical Methods, Computational Techniques and Intelligent Systems (MAMECTIS '16), Venice, Italy, January 29-31.*

30th International Conference on
Adaptive Structures and Technologies

# Investigation on a Morphing Winglet Driven by Differential Retractable Grid

**Wei Li[1*] and Ke Xiong[2]**

*[1]School of Air Transportation, Shanghai University of Engineering Science, 333#, Longteng Rd, Shanghai, China*

*[2]State Key Laboratory of Mechanics Control of Mechanical Structure, Nanjing University of Aeronautics and Astronautics, 29#, Yudao St, Nanjing, China*

*[*]rodway@163.com*

**Abstract**

The traditional winglet has the best drag reduction efficiency during the cruise phase. This paper presents a morphing winglet concept that utilizing differential retractable grid to change winglet height and cant angle during flight and optimize the drag characteristic of an aircraft at various flight conditions. The influence of morphing winglet on wake turbulence during takeoff phase (Ma=0.3, α=6° ) is validated by CFD and wind tunnel test. The results show that the wake turbulence of an aircraft and the drag coefficient has been reduced 56.6% and 5.9% by morphing winglet, compared to the traditional winglet.

**Introduction**

In recent years, a morphing winglet concept has emerged. It has a high efficiency of drag reduction during takeoff and cruise phase, through winglet deformation. Airbus and University of Bristol have jointly launched the "Morphlet" research project, which achieves active cant angle of winglet driving by servo motor (Bourdin, Gatto and Friswell, 2008; Gatto, Mattioni and Friswell, 2009). An active cant angle winglet is also investigated by University of Sao Paulo (Catalano and Ceron-Munoz, 2006). This paper presents a morphing winglet concept that utilizing differential retractable grid to change winglet height and cant angle during flight.

**Methods**

Fig. 1(a) shows the concept of differential retractable grid. The mechanism consists of a number of connecting rods and slider. The retractable grid 1 and the retractable grid 2 are respectively driven by two stepping motors. The upper rib and the slider are connected by a linear shaft guide rail, thereby ensuring that the slider can perform linear motion along the shaft guide rail and rotate around the shaft. In addition, the upper rib has a pair of lateral sliders that are slidable in a direction perpendicular to the chord line. The differential retractable grid is mounted on the wing as shown in Fig. 1(b).

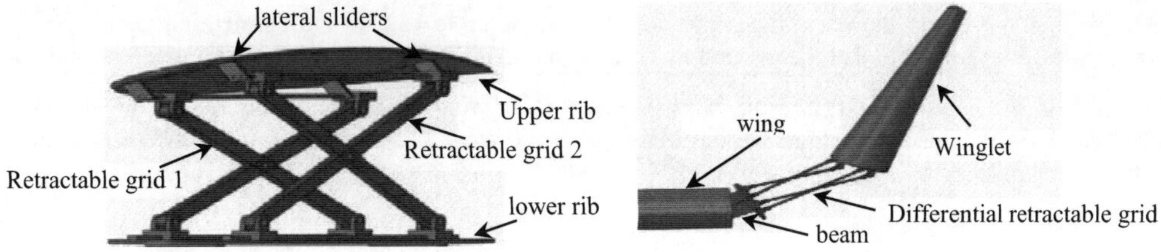


(a) Differential retractable grid (b) Mounting

**Figure 1**. Differential retractable grid

## CFD Simulation and Wind Tunnel Test

The influence of morphing winglet on wake turbulence during takeoff phase (Ma=0.3, α=6° ) is validated by numerical simulation and wind tunnel test. Taking the winglet of KC-135 aircraft as an example. According to the state before and after deformation of the morphing winglet, two winglet models with different height and cant angle have been established for CFD simulation and wind tunnel test, as shown in Fig. 2.

before deformation (H=0.135b/2, θ=20°) after deformation (H=0.163b/2, θ=29°)

**Figure 2**. CFD model

CFD and wind tunnel test result is shown in Fig. 3. Taking the 0.5c section behind the trailing edge of the wing as an example, the wake turbulence and the drag coefficient has been reduced by 56.6% and 5.9%.

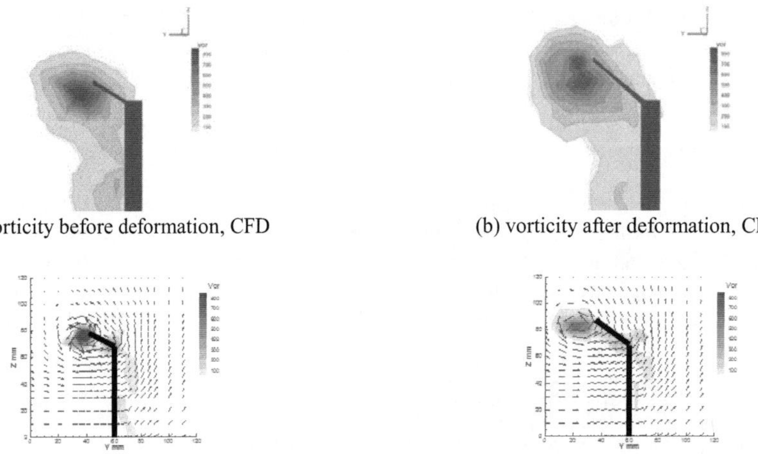

(a) vorticity before deformation, CFD (b) vorticity after deformation, CFD

(c) vorticity before deformation, wind tunnel test (d) vorticity after deformation, wind tunnel test

**Figure 3**. Comparison between CFD and wind tunnel test

## Results and Discussions

(1) Dynamic simulation results show that the expansion rate can be up to 13.9%, and the maximum change in the cant angle is 48.6°. Model test displays that the maximum expansion rate is 12.1%, and the change in the cant angle is 44°.

(2) CFD simulation and wind tunnel test results show that morphing winglet can effectively reduce the wake turbulence of aircraft. Taking the 0.5c section behind the trailing edge of the wing as an example, wake turbulence and the drag coefficient has been reduced by 56.6% and 5.9%.

## References


*Bourdin P, Gatto A and Friswell M.I (2008) Aircraft control via variable cant-angle winglets. Journal of Aircraft 45(2): 414-423.*

*Catalano. F.M, Ceron-Munoz. H.D (2006) Experimental analysis of aerodynamics characteristics adaptive of multi-winglets. Journal of Aerospace Engineering 220(3): 209-215.*

*Gatto. A, Mattioni. F, and Friswell. M.I (2009) Experimental investigation of bistable winglets to enhance wing lift takeoff capability. Journal of Aircraft 46(2): 647-655.*

# Analysis of Cylindrical Tensegrities for Use in a Morphing Boom

**Kaila Roffman[*] and George A. Lesieutre[**]**

*The Pennsylvania State University, University Park, Pennsylvania, 16802*

*[*]kmr88@psu.edu , [**]gal4@psu.edu*

**Abstract**

This research investigates the optimal design characteristics of a class-2 triplex tower as a morphing boom. The topic is to compare how changing the height of the nominal straight configuration influences the limits of achievable shapes. The tower is allowed to change overall shape through actuation of the supporting cables in each bay. With struts and face cables of fixed lengths, the range of achievable equilibrium shapes for a given structure is considered. We further determine which configurations may be reached from an initial straight configuration without member collision. For towers of a given overall height, we investigate the trade-offs between the number of bays and the height of single bays, with respect to the range of achievable shapes and complexity. Actuation methods are also considered. The results provide insight into the optimal characteristics for a triplex tower with a large range of achievable shapes, considering design decisions regarding the height of single bays, the use of multiple bays, and actuator placement.

**Introduction**

Tensegrity structures are self-equilibrated structures comprised of struts (compression members) and cables (tension members) connected only at their ends (nodes). Tensegrities considered in this study are cylindrical, indicating that the face planes maintain a fixed radius. Such tensegrities can be characterized by their n-plex and their class-k. "n" refers to the number of struts in a single bay (or nodes on the top / bottom faces). "k" is the maximum number of struts that meet at a single node. This research focuses on tensegrities that use cable actuation to change overall shape. A triplex (3-plex) is considered. Only supporting cables are allowed to change length, so struts and face cables have uniform fixed lengths. These lengths are determined geometrically by choosing a slant height (the distance from the center of the bottom face to the center of the top face) in the parallel configuration, a circumscribing diameter, and maintaining the required 600 twist angle between the bottom and top faces for a triplex in the parallel configuration.

**Approach**

*Equilibrated solutions for a single-bay triplex*. A geometry-based approach was used to determine possible configurations. For illustration, consider a single bay, cylindrical triplex. The bottom face is held fixed in space. All geometrically-possible shapes are found by considering spheres defined with the centers at the bottom nodes, with radius equal to the strut length. The surfaces of the spheres corresponds to possible end node locations, restricted further by the requirement that each end node be precisely one face-cable length from the others. Once all geometrically-viable shapes are found, the solution set is refined by keeping only those configurations which are in equilibrium. For a tensegrity in self-equilibrium, the struts must also all be in compression, and the cables must all be in tension.

The set of equilibrated shapes can be represented partially by the locations of the center of the top face in spherical coordinates. The results for a triplex of slant height 1 are shown in Fig. 2. For any given triplex, there is a minimum elevation angle above which there is an equilibrated solution for any azimuth angle. This minimum elevation is directly related to the height of the structure. The minimum elevation angle is determined as a function of the ratio of parallel slant height to circumscribing diameter. Increasing this non-dimensional height decreases the minimum elevation (results in a larger angular range of motion).

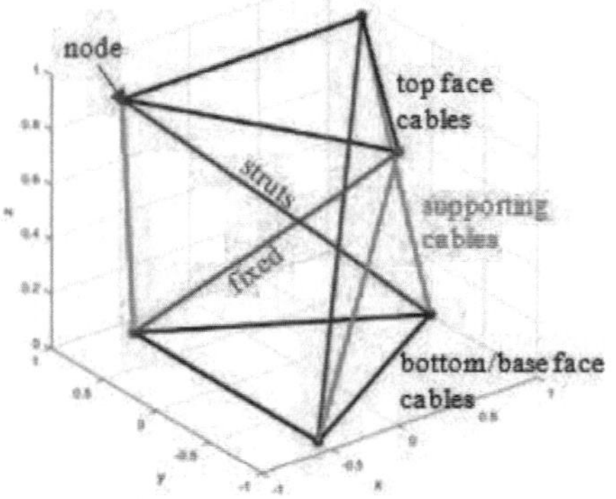


**Figure 1:** A cylindrical triplex of slant-height 1 in the parallel configuration

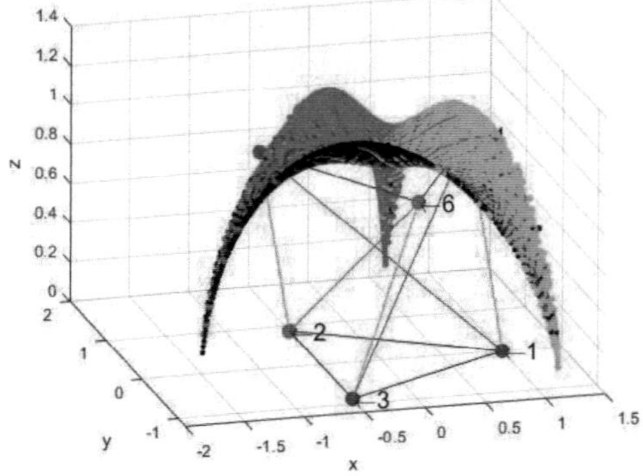


**Figure 2:** Centers of all equilibrated top faces plotted with the triplex in parallel configuration (slant height of 1)

*Multi-bay class-2 triplex structures.* Increasing the total height of a given tensegrity may be achieved by increasing strut lengths in a single bay, or by adding bays. Bays are mirror-stacked in a class-2 configuration. Ongoing research is exploring how additional bays increase the range of possible shapes for a triplex structure, primarily by considering the reachable volume of the center of the top-most face.

*Member collision & path finding.* The minimum elevation at which any azimuth provides a shape in equilibrium does not guarantee that it will be reachable from a starting configuration without member interference. With the parallel configuration as the starting configuration, this research explores potential paths a single bay triplex can take to different end shapes without member interference. This is also applied to multi-bay structures to consider the optimal conditions for maximizing range of achievable shapes.

*Actuation.* This research examines achievable shapes for a cable-actuated tensegrity. Potential actuation methods are considered and explored for their feasibility of use with tensegrities. Alternative actuation schemes, required cable-actuation lengths, and mass assessments are being addressed.

*Structural properties.* In addition to the geometric analysis of triplex configurations, this study also addresses structural properties such as stiffness and packaging efficiency. Bending and torsional stiffness are considered in relation to the slant height for comparable single and multi-bay tensegrities. These results will provide insight regarding the potential use of tensegrity structures as booms in space applications.

**Results and Discussion**

This research pursues the design analysis of cable-actuated cylindrical triplex tensegrities. A set of equilibrated solutions is found for a single bay, those reachable from an initial straight configuration are identified, then extension to multi-bay structures are considered. The range of motion generally increases with increasing ratio of slant height to circumscribing radius. Furthermore, the use of multiple bays greatly increases the range of motion. This research will also address actuation methods, including their influence on the range of reachable shapes (particularly in multi-bay cases). The relation between slant height and structural properties such as stiffness and packaging efficiency will also be described. An optimal solution considering design characteristics such as bay height and number of bays for a given tower height to maximize range of achievable shapes will be considered.

# Conceptual Design for the Actuation and Skin Integration of a Morphing Leading Edge

**Amin Fereidooni*, Jan Marchwica, Natalie Leung, Jaye Mangione, and Viresh Wickramasinghe**

*National Research Council of Canada, Ottawa, K1V 2B1, ON, Canada*

**Amin.Fereidooni@nrc-cnrc.gc.ca*

**Abstract**

Leading-edge high-lift devices increase the angle-of-attack at which the wing stalls. This allows for a greater maximum lift coefficient, and thus more lift for takeoff and landing. The three-position slat is the most common leading-edge high-lift device in use on commercial airliners today. Three-position slats have two major limitations: high noise levels, and increased drag. The slot opened during landing creates high noise levels which, due to noise regulations, can limit an aircraft's available approach paths and takeoff/landing hours[1]. In the takeoff position, small discontinuities and leakage of air from the bottom to the top of the wing can increase drag, which is undesirable. These limitations can be mitigated by using a morphing wing design. Current morphing wings are adapted to bend to create a high-lift profile. A flexible leading or trailing edge is bent downward to increase the lift of the wing. One example of a leading-edge morphing wing design is the DLR SADE design[2]. In this design, a flexible composite skin structure surrounds a rigid actuation mechanism[2]. The front 9% of the wing is constructed using the flexible skin, which is discretely-supported at four points by the rigid mechanism. Another example of a bending, morphing, high-lift device is the Flexsys Flexfoil[3]. This design uses a proprietary compliant actuation structure to change the shape of the wing[3]. A collaboration between NASA and the Air Force Flight Research Laboratory was initiated to flight-test the Flexsys Flexfoil morphing trailing edge design, in order to assess the real-world performance of a morphing wing[3]. The trailing-edge Fowler flap of a Gulfstream jet was replaced by the variable-geometry morphing surface developed by Flexsys. The device was successfully flight-tested from -2° to +30° deflection[3]. Although the aforementioned designs have been successful in many regards, they are limited by several factors. The DLR SADE design successfully demonstrated that a flexible leading edge could be optimized to provide reasonable fidelity to a target shape. However, the design suffered from poor shape control at varied angles of attack[2]. In this type of design, the skin thickness and support structure are optimized to provide the desired shape only at selected angles of attack[2]. The varying aerodynamic forces at other angles create undesired deformations in the skin, increasing the amount of deviation from the optimal shape. This is an issue with many morphing wings, since most current morphing designs for high-lift devices have limited mobility. A morphing wing concept is being developed by the National Research Council of Canada (NRC) with the purpose of creating a seamless high-lift device for noise and drag reduction. This project aims to develop a conceptual design which overcomes some of the challenges encountered with the SADE and Flexfoil designs. The NRC project explores the use of an elastomeric skin material and more flexible support structure to design a morphing wing with improved mobility. Initial validation of the morphing wing design performed using finite-element computer simulations will be presented. This paper focuses on the work performed to develop the optimal airfoil shape, and to analyze the integration of the skin with the flexible support structure. The design will allow an increase in chord length, thereby overcoming a performance limitation of existing morphing wing designs.

**Sample Results**

**Aerodynamic Optimization**

To create the drooped-nose profile, the wing was initially assumed to be flexible from 0% to 20% chord. The wing thickness and radius of curvature were allowed to change to provide the maximum suction on the upper surface. The NRC Aerodynamics Laboratory generated several optimized airfoil shapes by applying varying levels of stretch to the curve length forward of 20% chord. Several shapes were produced by varying the skin stretch from 0% to 20%; these are presented in Fig. 1.

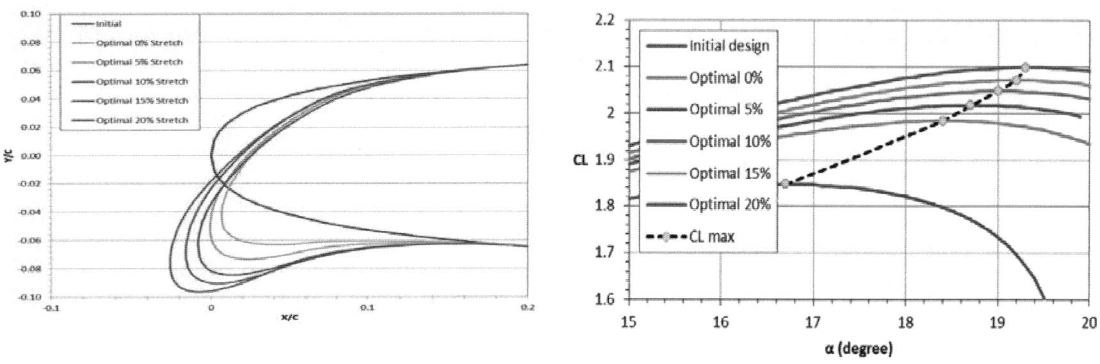


(a) (b)

**Figure 1.** (a) Airfoil shapes for different amounts of stretch and (b) the associated lift coefficients

## Deformation under Aerodynamic Loading

A structure was required to support the skin, in order to limit deformation under aerodynamic loads. This structure needed to provide minimal resistance in the chord-wise direction, to allow stretching of the skin. The structure also needed to constrain the skin from moving as a result of the aerodynamic suction forces. An example of such structure is a flexible accordion-style structure to support the skin. To determine an appropriate skin thickness and the distance between the supported points (the accordion bars), the effect of the aerodynamic pressure load on the skin was modelled. Since the skin is constrained discretely by the accordion structure, the pressure load causes periodic "bumps" in the skin surface between the accordion bars, shown schematically in Fig. 2. To determine the height of these bumps, a finite-element method (FEM) model of the skin segment between two adjacent accordion bars was modelled in MSC® Marc®. The maximum pressure was applied, and the length was varied to simulate different accordion support spacing options. The maximum skin deflection at the span-wise centerline was recorded for various spacings, with a span of 30 cm and skin thicknesses of 0.5 mm, 1 mm and 2 mm.

**Figure 2**. Finite Element Model of the Skin

## References


[1]*Shmilovich, A., Yadlin, Y. and Pitera, D. M., "Wing Leading Edge Concepts for Noise Reduction," 27th International Congress of the Aeronautical Sciences, ICAS 2010-5.2.2, Vol. 5, ICAS, Germany, 2010.*

[2]*Kintscher, M., Wiedemann, M., Monner, H. P., Heintze, O., and Kühn, T., "Design of a smart leading edge device for low speed wind tunnel tests in the European project SADE," International Journal of Structural Integrity [online database], vol. 2, no. 4, pp. 383-405, 2011, URL: http://dx.doi.org/10.1108/17579861111183911 [cited 17 October 2016].*

[3]*Kota, S., "Flight Testing of the FlexFoil Adaptive Compliant Trailing Edge," 54th AIAA Aerospace Sciences Meeting [online database], AIAA 2016-0036, 2016 URL: http://dx.doi.org/10.2514/6.2016-0036 [cited 17 October 2016].*

# A Morphing Winglet Concept with a Variable Stiffness Skin

**Jian Sun[1,2*], Linzhe Du[3], Fabrizio Scarpa[2], Yanju Liu[3], and Jinsong Leng[1]**

[1]*Center for Composite Materials and Structures, Harbin Institute of Technology, Harbin, China*

[2]*Advanced Composites Centre for Innovation and Science (ACCIS), University of Bristol, Bristol, UK*

[3]*Department of Astronautical Science and Mechanics, Harbin Institute of Technology, China*

[*]*sunjian@hit.edu.cn*

**Abstract**

Morphing wingtip technology could improve the reduction of the induced drag, fuel consumption, and take-off and landing distances compared with fixed wing. This work presents a morphing winglet concept based on active inflatable honeycombs and Shape Memory Polymer Composite (SMPC) skins, which could allow the winglet deploy with a morphing skin able to withstand the aerodynamic loading with distributed pneumatic muscle fibers. An actuation geometric and mechanical model of the morphing wingtip concept is developed and analyzed. A morphing wingtip prototype is also fabricated to demonstrate the kinematics and actuation. Experiments and simulations show agreement about the output angle/input pressure relationship necessary for the wingtip deployment.

**Introduction**

This article presents a new kind of variable stiffness skin made of Shape Memory Polymer Composite (SMPC), which can be hard in and convert to high elasticity in rubber state. Emulation and experiment were taken for this design. Pneumatic muscle fibers were settled skin to avoid broken when the skin is in rubber state. Emulation and experiment were also taken for this structure. The results show the design is feasible and the pneumatic muscle fibers (PMF) are sure to enforce the structure.

The skin contains two parts. One is the SMPC. Traditional SMP is brittle so it cannot transform too much and life span is also short. Spandex fibers were added into SMP to improve its mechanical property. The new SMPC shows good performance as supposed. Another part is PMF, it was settled under the polymer. When the SMPC is in glassy state, the PMF doesn't work and when the SMPC turns to rubber state, the PMF was inflated to support the SMPC via direct bonding and avoid failure.

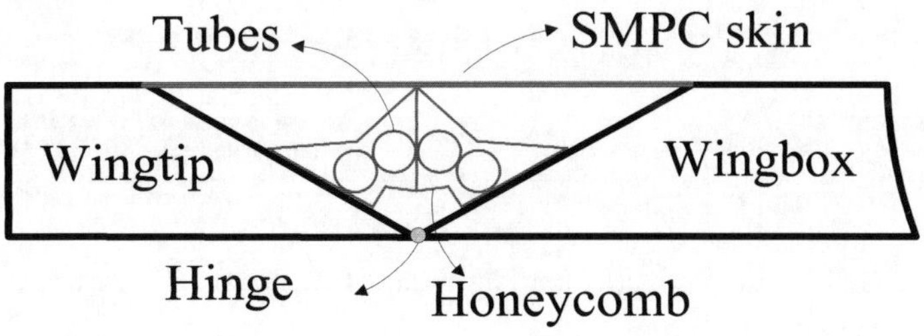


**Figure 1**. SMPC maximum displacements with/without PMF stringers at different pressures

**Methods/Model Formulations**

- Model analyses

Mechanical analyses were taken to this model and the recovery force was given as below:

$$F_1 = \left(F_{t1} - F_{t0}\right) = \frac{EI}{\cos\left(\frac{\theta}{4}\right)}\left(\frac{1}{\left(r_1 - \frac{t}{2}\right)^2} - \frac{1}{\left(R_0 - \frac{t}{2}\right)^2}\right) \quad (1)$$

- FEM simulations

  FEM simulations were taken to verify the effects of PMF. The results are shown in Table.1, which shows that PMF could enhance the rigidity of SMPC skin by one order of magnitude.

- Experiments

  A bending test was taken to get the modulus of PMF. And a verify experiment was taken to obtain the load bearing capacity of the SMPCs with and without pneumatic muscle fibers. Water pressure was used to impersonate air load, and the results are shown in Fig.2, which shows PMF could enhance the rigidity of SMPC about 100%. Furthermore, a simple model was made and a morphing action was taken.

**Results and Discussions**

The simulations and experiments show agreement to this variable stiffness tube reinforced shape memory composite skin. Further research was suggested in the future.

**Table 1**. Maximum displacements of the SMPC system in different states

| State | Maximum displacement [mm] |
|---|---|
| Glass state without PMF | 1.646e+00 |
| Glass state with PMF | 2.087e-01 |
| Rubber state without PMF | 3.292e+02 |
| Rubber state with PMF | 3.227e+00 |

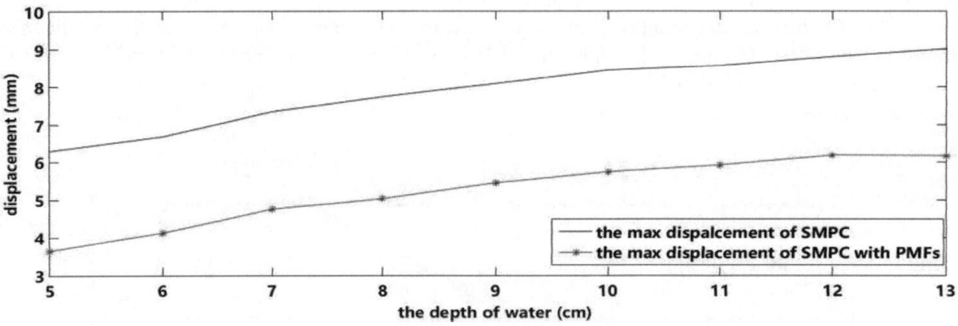


**Figure 2**. SMPC maximum displacements with/without PMF stringers at different pressures

# Surrogate Model of the Aerodynamic, Structural and Mass Properties of a Shape Adaptive Airfoil for the Blade of the Bo105 Helicopter

**Stephane Fournier*** **and Benjamin K. S. Woods**

*Bristol Composites Institute (ACCIS), University of Bristol, Bristol, UK*

**stephane.fournier@bristol.ac.uk*

**Abstract**

This paper introduces a fluid structure interaction analysis based surrogate model of the Fish Bone Active Camber (FishBAC) morphing aerofoil concept. One of the most challenging aspects of integrating morphing (or adaptive) aerostructures into real world applications is the difficulties presented by the need to capture the physics of the morphing device at a component level when considering the impact of the morphing device on the performance of the overall system. This surrogate model introduced here is intended to simplify this process by creating a reduced order model of the morphing component that can be run at very low computational cost within a helicopter rotor level analysis tool - while still capturing the coupling between the aerodynamics, structures and actuation that are inherent to compliance based morphing devices like the FishBAC. This paper will present the development of this surrogate model, explain the analysis methods used and the parameter space explored.

**Introduction**

Active camber morphing airfoils are one of the most promising concepts to potentially improve the aerodynamic behavior and reduce the weight of wings and blades mainly by replacing flaps, due to the benefits of a continuous shape variation. The applicability of the Fish Bone Active Camber concept (FishBAC, Figure 1a) to the blade of an MBB Bo105 helicopter is currently investigated within the European Commission funded Horizon 2020 research program called Shape Adaptive Blades for Rotorcraft Efficiency (SABRE). The goal is to provide a wide range of controllable lift coefficients and pitching moments with low drag penalty at actuation frequencies of up to 2/rev (14Hz). This would allow for optimization of blade twist between flight regimes and to better balance the thrust generation across the rotor disc to reduce induced losses. Previous numerical simulations confirm a high control authority of the concept over lift and pitching-moment coefficient and high levels of aerodynamic efficiency over a wide range of operating conditions and morphing deflections (Fournier S, Woods BKS, 2019). Full-scale rotor comprehensive analysis run using CAMRAD II showed gains in hover Figure of Merit of up to 5.5% (static morphing), and up to 5% in forward flight when dynamically actuated at 2/rev (Rauleder J, van der Wall BG, Abdelmoula A, et al., 2018).

**Methods/Model Formulations**

An improved Fluid-Structure-Actuator Interaction analysis has been developed which couples the MSES Euler CFD code (with boundary layer corrections) to a Beam Theory based structural model and a linear elastic actuation tendon model. It is able to predict the steady aeroelastic behavior of the device. The code allows for calculation of the deformed equilibrium shape, aerodynamic coefficients and actuation requirements of the FishBAC under quasi-static aerodynamic loading. The aerodynamic solver used is suitable for the analysis of high Mach numbers flows which is required to cover the whole range of operating condition at which the morphing rotor blades operate.

A surrogate model based on a full-factorial Design of Experiment (DoE) was generated to predict the steady aeroelastic performance of the FishBAC using MATLAB. The range of flight conditions endured by the Bo 105's rotor defines the range of local operating conditions along the blade at which the shape adaptive airfoil must operate and is listed in Table 1. The effect of the actuation input is investigated for both upward (-6°) and downward (+30°) trailing edge deflections (positive downward). Of the 4,332 total FSI runs, slightly less than 32% achieved convergence. Most of non-converged cases correspond to large angles of attack and large morphing deflections at high Mach numbers.

The cross-sectional properties of the baseline blade section have been computed and compared with the morphing blade section. To do so the BECAS software has been used and the output adapted to describe the shape adaptive blade (Figure 1b).

**Table 1**. Input parameters and ranges of the surrogate model

| input parameters | value range | samples |
|---|---|---|
| $\alpha_e$ | [-6 :1: +12] ° | 19 |
| $\delta_F$ | [-6 :2: +30] ° | 19 |
| $M_\infty$ | [0.12 :0.06: 0.78] | |
| $V_\infty$ | [40.84 :20.42: 265.43] m s$^{-1}$ | 12 |
| $Re_\infty$ | [0.7454 :0.3727: 4.8449] · $10^6$ | |

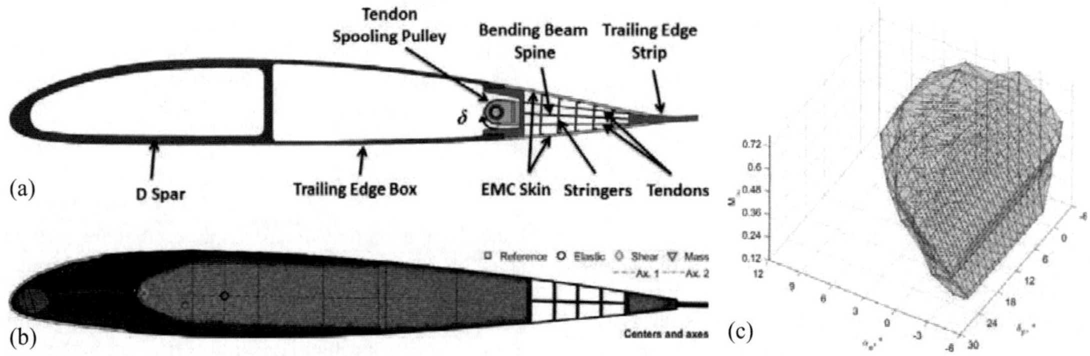


**Figure 1**. (a) FishBAC morphing concept, (b) Bo-105 blade with FishBAC schematic, mesh and properties overlaid and (c) converged interpolation domain boundary in the 3D design space of the surrogate model.

### Results and Discussions

The FSI predicted trailing edge tip deflections ranging between −2.2% and +7.9% of the chord length with the device requiring a maximum applied torque per unit span length ranging from 8.6 Nm (upward defl.) to 39.5 Nm (downward defl.). The MSES solver, while limited in the range of operating conditions it would converge for (Figure 1c), was generally able to capture performance before separation occurs and also captured the increased drag when shocks occur – which previous analysis was unable to capture.

Most of the mass and stiffness properties of the Baseline blade computed with BECAS are retained well with errors lower than few percentage points when the FishBAC is integrated, with a maximum error of 5.87% for the torsional term GJ. The changes in the cross section produce a discrepancy in the lag stiffness term which is more than double the baseline value and should be carefully considered in the blade design.

### References

*Fournier S, Woods BKS (2019) Surrogate Modelling of the Steady Aeroelastic Performance of the Fish Bone Active Camber Concept. AIAA Scitech 2019 Forum, AIAA SciTech Forum, (AIAA 2019-1859), doi: 10.2514/6.2019-1859*

*Rauleder J, van der Wall BG, Abdelmoula A, et al. (2018) Aerodynamic Performance of Morphing Blades and Rotor Systems. Proceedings of the AHS Forum, 14-17 May 2018, Phoenix, Arizona*

30th International Conference on Adaptive Structures and Technologies

# A Finite Element Analysis of a Gull Wing through Varied Morphed Configurations

**Christina Harvey*** **and Daniel J. Inman**

*Department of Aerospace Engineering, University of Michigan, 1320 Beal Ave, Ann Arbor MI, USA*

*harveyca@umich.edu*

## Introduction

Throughout the history of aviation, birds have provided continued inspiration for aircraft design. This is in part because birds exhibit impressive maneuverability and control when compared to current aircraft. This skill is often attributed to a bird's ability to dynamically change the shape of their wings, which allows them to modulate performance and stability requirements thus increasing a bird's operational range (Lentink et al., 2007) . Implementing an avian-inspired structure on an aircraft is non-trivial due to the wing's complex structure which is comprised of a skeleton, skin and feathers (Altshuler et al., 2015). Recent advances in smart materials and adaptive structures have made the design of a bio-inspired morphing wing a possibility (Barbarino et al., 2011). However, before a bio-inspired wing can be implemented on an aircraft, it will be necessary to understand how the stress and deformation states vary throughout different morphed configurations while under aerodynamic loading. In this study, we investigate the stress and deformation state of three major bones: the humerus, ulna, and carpometacarpus for a glaucous-winged gull (Larus glaucescens) in gliding flight, and how these states vary with elbow and wrist extension.

## Methods/Model Formulations

The wing shapes used in this analysis are taken from a previous anatomical study where we identified the 3D positioning of the skeletal system and the corresponding wing outline throughout a bird's range of motion (Harvey, Baliga, Altshuler, et al., 2019). From this dataset we selected a subset of the wings that correspond to an extended wing configuration. This configuration has previously been shown to be commonly used by gulls while gliding (Harvey, Baliga, Lavoie, et al., 2019). With the 3D coordinates of the skeleton and wing we constructed models of the wings in Solidworks. Next, the wing models were imported into ABAQUS, a software for finite element analysis. The models were imported as 3D, solid, homogenous, shell structures. Material properties were assigned to three distinct regions: feathers, muscles, and bones. All materials were modelled as isotropic which is a large simplification of the true material properties but still provides a useful approximation of the expected structural response. The feather and muscle shell regions were meshed with a linear 4-node quadrilateral (S4R) or a linear 3-node triangular (S3R) elements using a quad-dominated meshing command. The bones were represented by beams and meshed with 2-node linear beam (B31) elements.

The wings were loaded using force and moment distributions that were previously determined using a numerical lifting line method (Harvey, Baliga, Altshuler, et al., 2019). The loads were determined for a flight speed of 19.9 m/s which is at the upper range of the observed gliding speed for gulls (Shepard et al., 2016). The lifting line results provide an estimation of the lift and drag distribution on the wing, as well as a pitching moment about the quarter chord. We implemented the loads from the lifting line study as concentrated loads applied at 40 control points at the quarter-chord along the span of the wing. The wing was modelled as if the bird was providing enough muscular stabilization to maintain the shoulder in a fixed position. This was implemented by fixing all translational and rotational degrees of freedom of the node at the shoulder joint. In addition, the ligament and muscles attaching to the humerus were modelled as a pinned connection. This was done by fixing all three translational degrees of freedom of the edges extending from the leading edge point to the shoulder joint and from the shoulder joint to the root of the wing. We assigned a displacement boundary condition along the wing root edges located behind the humerus, constraining motion in the y-direction. This boundary condition effectively simulates feathers that can move in the x- and z-direction but not into or away from the body along the y-axis.

## Results and Discussions

Our results indicate that morphing the elbow and wrist affects the stress and deformation states for the three major bones of an avian wing. We found evidence that the gull wing is largely over-designed for steady gliding flight. This result suggests that weight or construction costs could be reduced by using a material with a lower ultimate strength than bones due to the large safety factors. Our analysis also revealed that extending the elbow and wrist tended to reduce the maximum Von Mises stress in the bones, except for the humerus where extending the elbow increased the stress. In addition, we found that extending the wrist and elbow had opposite effects on the location of maximum stress and the magnitude of maximum deformation in the bones. In all, the results suggest specific structural reinforcement strategies that can be used to inform future bio-inspired morphing wing designs.

## References

*Altshuler DL, Bahlman JW, Dakin R, et al. (2015) The biophysics of bird flight: functional relationships integrate aerodynamics, morphology, kinematics, muscles, and sensors. Canadian Journal of Zoology 93: 961–975.*

*Barbarino S, Bilgen O, Ajaj RM, et al. (2011) A review of morphing aircraft. Journal of Intelligent Material Systems and Structures 22: 823–877. DOI: 10.1177/1045389x11414084.*

*Harvey C, Baliga VB, Altshuler DL, et al. (2019) Pitch Control Effectiveness of the Avian Elbow and Wrist via a Numerical Lifting Line Analysis. In: AIAA Scitech 2019 Forum, 6 January 2019. AIAA SciTech Forum. American Institute of Aeronautics and Astronautics. DOI: 10.2514/6.2019-0853.*

*Harvey C, Baliga VB, Lavoie P, et al. (2019) Wing morphing allows gulls to modulate static pitch stability during gliding. Journal of The Royal Society Interface 16(150): 20180641. DOI: 10.1098/rsif.2018.0641.*

*Lentink D, Müller UK, Stamhuis EJ, et al. (2007) How swifts control their glide performance with morphing wings. Nature 446: 1082.*

*Shepard ELC, Williamson C and Windsor SP (2016) Fine-scale flight strategies of gulls in urban airflows indicate risk and reward in city living. Philosophical Transactions of the Royal Society B: Biological Sciences 371: 20150394. DOI: 10.1098/rstb.2015.0394.*

30th International Conference on Adaptive Structures and Technologies

# Analysis, Simulation and Comparison of an Exoskeleton for Rehabilitation of Lower Limbs using Different Materials for Its Manufacture

C. Hernández-Santos[1], Fermín C. Aragón[2*], Adriana Vargas-Martinez[2], Jorge de J. Lozoya-Santos[2], Ricardo Ramírez-Mendoza[2], and Juan C. Tudón-Martínez[1]

[1] *Universidad de Monterrey, Monterrey, México*

[2] *Tecnologico de Monterrey, Monterrey, School of Science and Engineering, México*

**A00825397@itesm.mx*

**Abstract**

This work is the continuation of the development and preliminary manufacture of an exoskeleton for lower extremities with 5 degrees of freedom, from the established design parameters the prototype was manufactured, an analysis of the finite element was made using ANSYS mechanical and Solidworks for the mechanical design with the purpose of studying the mechanical behavior of the structure when subjected to stresses and deformations. Within the simulations, another type of material was exchanged so that the future manufacture is made with lighter materials and with the possibility of being manufactured using 3D printing, the materials proposed for this experimentation were PLA , ABS and PLA with carbon fiber the mechanical behavior of these materials was compared with that of an exoskeleton currently manufactured with aluminum.

## Introduction

In Mexico during 2010 was recorded that 5,739,270 people who have disability, representing 5.1 % of the total population. The limitation of mobility is the greatest suffering among the population; 58.3 % of the declared limitations relate to walk or move. One of the diseases that affect mobility is walking paraplegia, which affects more than 20 million people in the world, where these people have no or partial capacity for mobility in their lower extremities. (INEGI, 2013). The growing number of people with motor disabilities march has been engineering to develop new techniques to assist walking. One of the techniques under development are driving exoskeletons. These devices are intended to give patients the possibility to regain mobility and thus actively integrate into society, overcoming physical limitations associated with the inability to walk. Exoskeletons are a technological alternative to wheelchairs or passive orthosis in which the patient performs support and supported rolling walkers, crutches or canes. Because of the above, this paper presents a new exoskeleton proposal for the rehabilitation of people suffering from partial paraplegia, i.e. those who have lost mobility and strength in one of their lower limbs as a result of an accident, but still have some control of their extremity. The work presents the development of the prototype, such as the inverse and forward kinematics analysis, besides the Lagrange dynamics and the development of a power circuit for the exoskeleton control. mechanism design and optimization, kinematics analysis, trajectory planning, dynamic analysis, motion simulation, and experiments for a new lower limb exoskeleton.

## Methodology

The design of an exoskeleton is a subject of high complexity due to the different considerations and issues exist simultaneously. The proposed prototype for exoskeleton is developed by CAD software as shown in Figure 1. The exoskeleton has a five DOF and a range of movement equivalent to that of human leg with restrictions on the movements of flexion, extension, abduction and adduction movements for the lower limb.

L1, L2 and L3 shown in Figure 1 are the main lengths of the exoskeleton in the lateral plane, the dimensional detail of the lengths is shown in Table 1 where the units are expressed in meters, the first prototype constructed was made by way of Fixed, however as future work is proposed in the development of links that adapt to the dimensional lengths of users with different stature.

Table 1. Dimensions of lateral plane exoskeleton links

| Part of the exoskeleton | Dimension |
|---|---|
| Femur(L1) | 0.39 m |
| Tibia (L2) | 0.36 m |
| Ankle(L3) | 0.20 m |

Table 1 describes the lengths of the main exoskeleton links shown in Figure 1 in their side view, where each L is a link length which is expressed in centimeters. Figure 2 shows the proposed CAD rendering in the extended and flexed position, where Figure 2a is the flexed position and Figure 2b is the extended position.

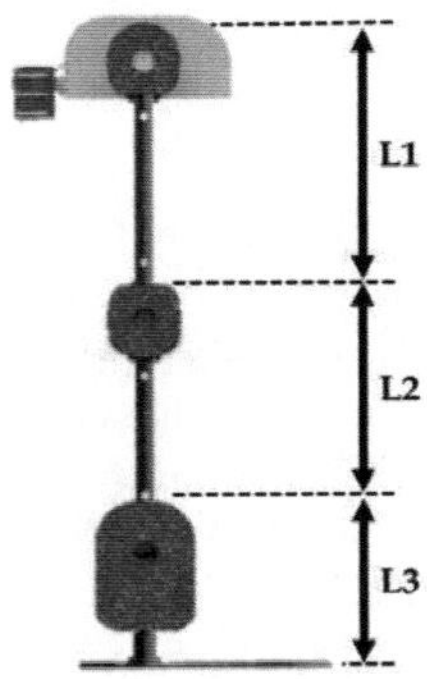


**Figure 1**. Preview of exoskeleton on lateral side

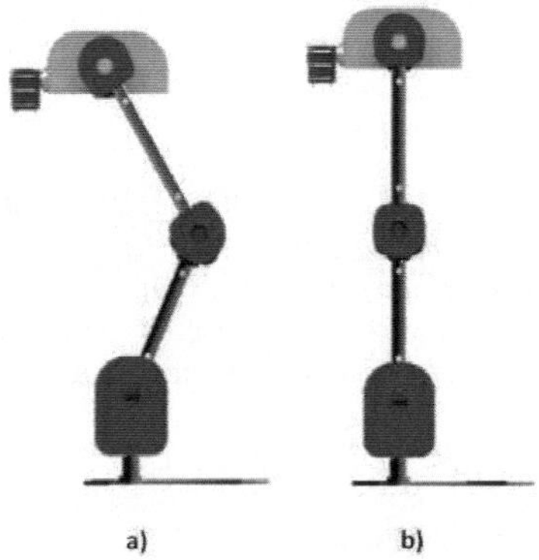


**Figure 2**. Exoskeleton designed by CAD software

## Conclusions

After having performed the simulations that determine the physical behavior of the exoskeleton and its manufacture has been performed, it is expected to perform performance tests to study its behavior and thus be able to implement a control system that adapts to the needs of the exoskeleton and the patient.

## References

*Aragón F.C., Hernández-Santos C., Hernández Vega JI., Córdova D.A., Palomares Gorham D.G., Sánchez Cuevas J.L. (2017) Design and Simulation of a New Lower Exoskeleton for Rehabilitation of Patients with Paraplegia. In: Sidorov G., Herrera-Alcántara O. (eds) Advances in Computational Intelligence. MICAI 2016. Lecture Notes in Computer Science, vol 10061. Springer, Cham*

*INEGI. (2013). Instituto Nacional de estadística y geografía. Las personas con discapacidad en México una visión al 2010, México.*

# Biodegradable Starch-based Materials for Wound Dressing, Monitoring, and Healing

**M. Mohsen Delavari[1*], A. Perez-Cruz[2], and I. Stiharu[1]**

[1]*Department of Mechanical, Industrial and Aerospace Engineering, Concordia University, Montreal, Quebec, Canada*

[2]*Autonomous University of Queretaro, Mexico*

[*]*mohammadmohsen.delavari@concordia.ca*

**Abstract**

About two decades ago the healthcare took a turn as the access to technologies enabled development of extremely enabling and accessible medical devices, However, problems related to the safe disposal of the used devices and materials surfaced later, when the access to such devices became very popular. Materials such as polystyrene, polyvinyl chloride, and nylon which are among the materials used in the fabrication of advanced medical devices are abundant and meanwhile, they require very long duration to decompose within the regular environment. Materials from syringes to surgical kits need to be transformed at a high environmental cost. Such polymer-based materials are used not only in medical technology but also in other mass-distributed products such as food recipients, food packaging device, and others. On the other hand, global warming is now one of the most concerning issues at all society levels around the world. Although the efforts to regulate the disposal of non-recycling materials have been made, the progress is still slow. These environmental regulations, social concerns, and growing environmental understanding throughout the world have triggered renewed efforts in varieties of industries to develop new products and processes compatible with our environment. The replacement of plastic material made parts or those fabricated from conventional cellulose with biodegradable starch-based papers and blends could represent a game changer in the environment protection action by reducing the usage of plastics and also by protecting the forest cutting. Starch is known as a native, renewable, abundant, low-cost and biodegradable material which is produced by agriculture plants such as corn, wheat grains, potato, rice, other cereal, etc. (Jeong et al., 2018). Starch is considered as a substitution to conventional materials. Same modified material type may also be considered in other useful applications in order to substitute expensive disposable environmental materials. The wound dressing is the one that will be discussed in the paper below. As a result of accidents, many hundreds of patients suffer from skin damage due to burns by hot water, flames accidents, and boiling oil every year. Due to medical concerns as well as the ecological safety concerns, natural and also biodegradable polymer wound dressings have recently attracted some attention (Kamoun et al., 2017). Starch-based customized biodegradable materials for general and functional applications could be developed for wound dressing, wound healing, and drug delivery systems. Moreover, for specific cases, ultraviolet (UV) protection and active status monitoring of the wound area could be inferred in order to protect the wound from microbial invasion, prevent strong local pain and infections and facilitate fast healing. As mentioned above, UV protection capability could be applied to this kind of blend, therefore, near UV, visible, and infrared wavelength FT-IR spectroscopy analysis of starch, polyvinyl alcohol (PVA), glycerol, and citric acid blends prepared with different amount of citric acid (1g and 2g) is performed and represented in Figure 1. The results show almost the same absorbance near UV and visible; however, the absorbance in the infrared spectrum is higher than the shorter wavelengths. Moreover, it could be seen that the blend with 1g citric acid in the receipe has more capability to absorb the near UV (350 nm) wavelength.

Active state monitoring devices directly embedded on the starch-based wound dressing could perform humidity, oxygen ($O_2$), and $CO_2$ measurement at the wound surface. Moreover, these materials could embed healing proteins that could accelerate the healing process (Sorg et al., 2017). Antibacterial property of starch-based material represents a feature to support this type of application, which is the wound dressing.

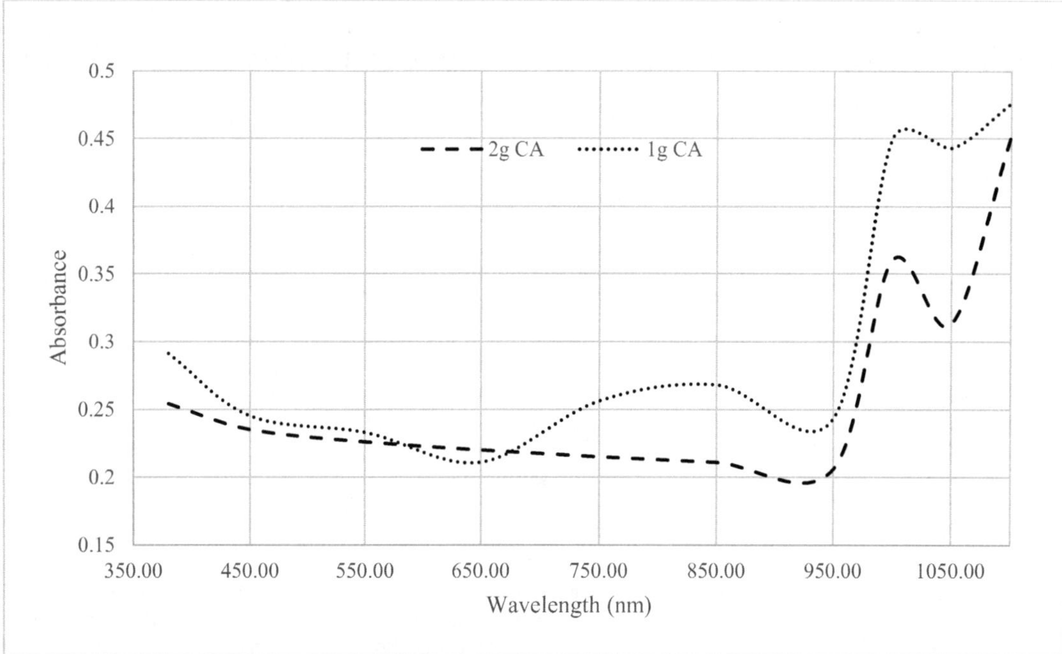


**Figure 1**. FT-IR spectra of starch, polyvinyl alcohol (PVA), glycerol, and citric acid (CA) composite films

The technology of producing starch-based materials, their properties and the possible variance of their properties with the variation of the concentration of the ingredients is presented in the literature [Jeong H, a.o., 2018]. The specifically targeted properties of the starch-based materials are also presented in other papers but they do not specifically address the aspect of UV absorbtion [Kamoun EA a.o., 2017].

Moreover, starched-based material may be built with transparent and flexible characteristics which represents a positive feature for wound dressing. Same starch-based materials could serve as a substrate for wearable electronic devices which include different devices such as transistors, radio frequency identification devices, sensors, and energy storage devices (Park and Lee, 2019).

## References

*Jeong H, Baek S, Han S, et al. (2018) Novel Eco-Friendly Starch Paper for Use in Flexible, Transparent, and Disposable Organic Electronics. Advanced Functional Materials 28: 1704433.*

*Kamoun EA, Kenawy E-RS and Chen X. (2017) A review on polymeric hydrogel membranes for wound dressing applications: PVA-based hydrogel dressings. Journal of advanced research 8: 217-233.*

*Park MJ and Lee JS. (2019) Foldable and Biodegradable Energy-Storage Devices on Copy Papers. Advanced Electronic Materials 5: 1800411.*

*Sorg H, Tilkorn DJ, Hager S, et al. (2017) Skin wound healing: an update on the current knowledge and concepts. European Surgical Research 58: 81-94.*

# Design of Bypass Rotary Vane Magneto Rheological Damper for Prosthetic Knee Application

**Radhe Shyam Saini Tak[1*], Hemantha Kumar[1], Sujatha Chandramohan[2], and Sujatha Srinivasan[2]**

[1]*Department of Mechanical Engineering, National Institute of Technology Karnataka, Surathkal, Mangalore, India.*

[2]*Department of Mechanical Engineering, Indian Institute of Technology Madras, Chennai, India.*

[*]*radhe.chandu@gmail.com*

**Abstract**

In this paper, a bypass rotary vane type magneto rheological (MR) damper suitable for prosthetic knee application is designed. The torque and angular velocity requirements of the normal human knee are used as design limits. In the proposed design, the rotary vane chamber and the MR valve are connected by hydraulic cables and ports and are designed separately. The rotary vane chamber is designed based on the cross-sectional size limits of the normal human knee, while the MR valve is designed with the objective of obtaining the maximum on-state damping torque, minimum weight, and minimum off-state torque.

## Introduction

Prosthetic devices working on MR brake applications involve many components and thus require utmost manufacturing precision while the ones working on MR dampers use commercially available off-the-shelf MR dampers with an exception of a few (Seid et al., 2018). In the present study, a bypass rotary vane type MR damper shown in Fig. 1 is designed for a prosthetic knee application. The static and rotary vanes divide the rotary chamber into two fluid regions. The static vane is fixed to the rotary chamber, while the rotary vane moves with the shaft. The fluid regions are connected using hydraulic cables and ports with an MR valve located in between.

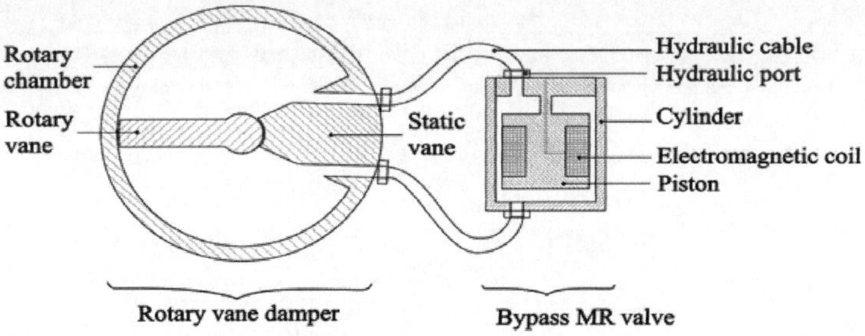


**Figure 1**. Rotary vane bypass MR damper

## Design Methodology

Neglecting the friction forces due to seals and the flow tube pressure drops, the damping torque of the rotary damper can be calculated using equation (1).

$$T_d = (\Delta P_\eta + \Delta P_\tau) A_{vane} R_{vane} \tag{1}$$

Here, $\Delta P_\eta$ is the pressure drop due to viscous forces in the MR valve, $\Delta P_\tau$ is the pressure drop due to the controllable shear forces, $A_{vane}$ is the cross-sectional area of the rotary vane and $R_{vane}$ is the distance between the centroid of the vane area cross-section and the rotary axis. For flow mode condition, the pressure drop due to viscous and controllable forces can be calculated using equations (2) and (3) (Yang et al., 2002),

$$\Delta P_{\eta} = \frac{12\eta Q L_t}{wh^3} \tag{2}$$

$$\Delta P_{\tau} = \left(2.07 + \frac{12Q\eta}{12Q\eta + 0.4wh^2\tau_y}\right)\frac{\tau_y L}{h} \tag{3}$$

Here, $w$ is the mean circumference of the annular flow path, $Q$ is the flow rate of the fluid in the gap, $h$ is the fluid gap, $L$ is the effective pole length, $L_t$ is the total axial length of the piston, $\eta$ is the apparent viscosity, and $\tau_y$ is the yield strength of the MR fluid. Neglecting the flow tube pressure drops, the flow rate in the vane chamber remains equal to the flow rate of fluid in the annular gap.

## Results and Discussions

An optimization problem is formulated considering six geometric variables of the MR valve and on-state torque, total damping torque and weight of the valve as objectives. A working speed of 8.4 rpm is used in the design. The Pareto optimal set obtained for the optimization problem is shown in Fig. 2. The optimal front points A – F shown in Fig. 2(a) show a trade-off in weight as shown in Fig. 2(b) while the points G-H show a trade-off in off-state damping torque.

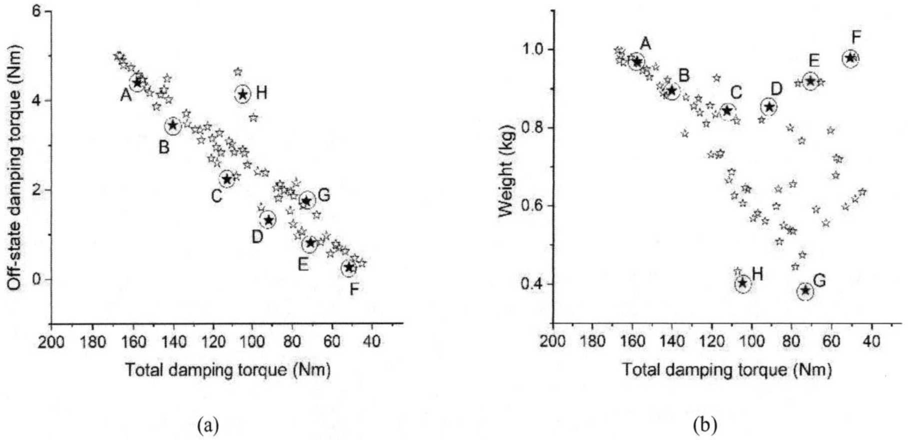


**Figure 2**. Pareto optimal set of (a) off-state and total damping torque and (b) weight and total damping torque.


## Acknowledgment

The Research was jointly supported by IMPRINT Project (Project No. IMPRINT/2016/7330) which is funded by Ministry of Human Resource Development and Ministry of Road Transport and Highways, Govt. of India and SPARC Project (Project no. 785) funded by Ministry of Human Resource Development.


## References


*Seid S, Chandramohan S and Sujatha S (2018) Optimal design of an MR damper valve for prosthetic knee application. Journal of Mechanical Science and Technology 32(6): 2959-2965.*

*Yang G, Spencer Jr BF, Carlson JD and Sain MK (2002) Large-scale MR fluid dampers: modeling and dynamic performance considerations. Engineering structures 24(3): 309-323.*

# Kinematic Analysis and Position Control of a Flexible Tendon-driven Catheter for Minimally Invasive Cardiac Surgery

**Mohammad Jolaei, Amir Hooshiar*, Javad Dargahi**

*Mechanical, Industrial, and Aerospace Engineering, Concordia University, Montreal, Canada*

**s_hooshi@encs.concordia.ca*

**Abstract**

To address the clinical need for force control of cardiac catheters in minimally invasive cardiac surgery, a tendon-driven flexible catheter was proposed. The proposed catheter was prototyped using silicon-rubber body and reinforced with a coil spring. Afterward, the catheter was modeled as a flexible continuum robot and its spatial workspace dome was obtained through kinematic analysis. Also, position control of the catheter was performed by establishing a nonlinear model relating tendon lengths and tip positions. Results exhibited fair agreement between the position trajectory and the desired path.

**Introduction**

Controlling the contact force between the tip of a cardiac catheter and cardiac wall is crucial (Hooshiar et al., 2019). Whereas, excessive force would puncture the cardiac wall and insufficient force would lead to an ineffective treatment for the patient. Previous studies have suggested two methods for controlling the contact force between the tip of a cardiac catheter and cardiac wall: image-based and sensor-based force control. These methods are prone to intrinsic limitations. In a previous study, (Jolaei et al., 2019, Hooshiar et al., 2017) proposed and validated a new force control framework using tendon-driven catheters. In continuation, the goal of this study was to propose and investigate the controllability of a novel tendon-driven flexible cardiac catheter. To this end, the spatial workspace of the propose catheter was kinematically analyzed, and a nonlinear relationship between the length of tendons and spatial position of the catheter was derived. Afterward, a straight path on the workspace was defined and catheter was driven with feedforward control. In the end, the accuracy of feedforward position control was analyzed.

**Material and Methods**

Fig. 1(a) depicts proposed catheter in its local coordination system. The local coordination system X-Y-Z was assumed to be at a fixed position with respect to the global coordination system. Also, the catheter was assumed to follow a constant radius bending while driven by the tendons. This assumption is previously validated in (Ganji et al., 2009). Since the length of the catheter remains constant throughout the bending, bending angle and bending radius would relate as:

$$L_0 = r_b.\theta_b \tag{1}$$

, where, $L_0$ was the length of catheter and $r_b$ and $\theta_b$ were bending radius and bending angle, respectively. Also, using the cosine law, the length of the chord formed by the catheter was derived as:

$$d = \sqrt{2}r_b\sqrt{1 - cos\theta_b} \tag{2}$$

Where, $\theta_b$was in the range of $[0 - \frac{\pi}{2}]$, and the rotation angle about longitudinal axis in the range of $[0 - \pi]$. Substituting $L_0$ in Eq. 1 and using Eq.2 in spherical coordination system, the spatial workspace of the catheter was found. Fig. 1(b) shows the theoretical workspace of the catheter. To obtain a relationship between the length of tendons and position of the catheter tip, Eq. 3 and Eq. 4 were used. Also, since the catheter was over-actuated, and tendons could only apply tensile force, only two tendons were used for position control of the catheter in each quadrant.

$$\begin{pmatrix} x \\ y \end{pmatrix} = \begin{pmatrix} a_{11} & a_{12} & a_{13} & a_{14} \\ a_{21} & a_{22} & a_{23} & a_{24} \end{pmatrix} \begin{pmatrix} L_1 \\ L_2 \\ L_3 \\ L_4 \end{pmatrix} + \begin{pmatrix} b_{11} & b_{12} & b_{13} & b_{14} \\ b_{21} & b_{22} & b_{23} & b_{24} \end{pmatrix} \begin{pmatrix} L_1^2 \\ L_2^2 \\ L_3^2 \\ L_4^2 \end{pmatrix} + \begin{pmatrix} c_{11} \\ c_{21} \end{pmatrix} \tag{3}$$

$$z = \begin{pmatrix} d_1 & d_2 & d_3 & d_4 & d_5 \end{pmatrix} \begin{pmatrix} x \\ y \\ x^2 \\ y^2 \\ 1 \end{pmatrix} \tag{4}$$

Where, $a_{11\text{-}24}$ and $b_{11\text{-}24}$ were the fitting parameters and $L_{1\text{-}4}$ were the length of tendons. To obtain the position of catheter tip two USB cameras (C920, Logitec Inc.) were calibrated with a chessboard template. Also tendons were driven using four independent stepper motors interfaced with a PC through and two independent Arduino Uno microcontrollers. Four desired paths were defined for the catheter to move along. The desired paths were defined as four sectors on four quadrants of the dome with equal lengths of 6mm. Fig. 1(a) shows the experimental setup schematically.

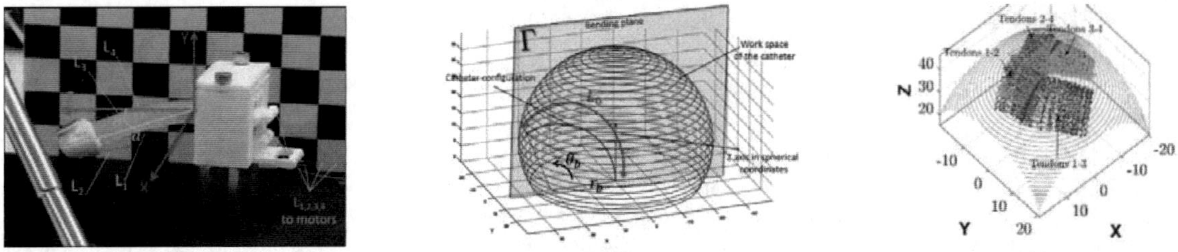


(a) (b) (c)

**Figure 1**. (a) Experimental setup (b) Theoretical workspace (c)

**Results and Discussions**

Fig.1(c) depicts the swept positions of the catheter tip and the theoretical workspace theoretically. The results showed that the kinematic analysis could accurately predict the spatial workspace (error<5%). Also, the results confirmed validity of the constant bending radius assumption. Error was defined as the absolute error between the desired position of the catheter and its actual position in each sector. The distribution of error in all the data normal with a mean of 0.5mm and maximum of 1.5mm.

**Conclusion**

This study was in continuation of previous efforts for position-based force control of a flexible cardiac catheter. The results showed validity of the kinematic analysis of the proposed catheter. Also, the feed-forward control strategy for driving the tendons was successful in fairly accurate position control of the catheter.

**References**


*Hooshiar, A., Najarian, S. and Dargahi, J., 2019. Haptic Telerobotic Cardiovascular Intervention: a Review of Approaches, Methods, and Future Perspectives. IEEE Reviews in Biomedical Engineering.*

*Jolaei, M., Hooshiar, A. and Dargahi, J., 2019, June. Displacement-based Model for Estimation of Contact Force Between RFA Catheter and Atrial Tissue with ex-vivo Validation. In 2019 IEEE International Symposium on Robotic and Sensors Environments (ROSE) (pp. 1-7). IEEE.*

*Hooshiar, A., Razban, M., Bandari, N.M. and Dargahi, J., 2017, June. Sensing principle for real-time characterization of viscoelasticity in the beating myocardial tissue. In 2017 IEEE International Conference on Computational Intelligence and Virtual Environments for Measurement Systems and Applications (CIVEMSA) (pp. 72-77). IEEE.*

*Ganji, Y. and Janabi-Sharifi, F., 2009. Catheter kinematics for intracardiac navigation. IEEE Transactions on Biomedical Engineering, 56(3), pp.621-632.*

# Fault Diagnosis of Adaptive Structures using Mahalanobis-Taguchi System

**Hiroshi Okubo*, Toui Ushiku, and Marika Satoh**

*Kanagawa Institute of Technology / 1030 Shimo-ogino, Atsugi, Kanagawa, Japan*

*okubo@me.kanagawa-it.ac.jp

**Abstract**

The Taguchi method (Mahalanobis-Taguchi system) is applied to the self-diagnosis problem, wherein a damage in a smart structure is detected using the observed vibration response data.

**Introduction**

In recent years, advanced fault detection technology is indispensable in space system development. Since human judgment of various diagnostic data is limited in complex systems such as spacecraft, it is necessary to introduce an autonomous fault detection system in order to prevent serious mission failures and system damages. One such method is Mahalanobis-Taguchi system (MT system), which has been used to date for fault detection of many complex systems. In this research, we focus on the MT system and apply it to the fault diagnosis of smart structures with advanced functions such as high precision shape control and vibration control. In this paper, we apply the MT system to the self-diagnosis problem that detects the failure in the smart structure using the random vibration response data.

**Methods/Model Formulations**

In the MT system, homogeneous group (basic and normal group) is defined, which serves as a unit space. The distance from the center of the unit space of the target data is defined as the Mahalanobis distance (MD), which is used as a measure of the extent of abnormality. In this study, fault diagnosis experiments were conducted on a laboratory model of smart beam. Diagnostic tests were performed using a cantilevered beam equipped with a piezo-ceramic actuator near the base of the beam. A laser displacement meter was used as the output sensor. The vibration response excited by a minute input can be monitored by using a sensor attached to the structure. This enables the system to self-diagnose and detect a failure or damage in the structure.

The diagnostic experiment was performed for an abnormality simulated by an artificial hole near the base of the beam (Figure 1). A random input was supplied to the smart beam and the MD test was applied to the measured time sequences of the beam displacements. The mode shapes of the beam vibration are shown in Figure 2. Figure 3 shows an example of the displacement responses of vibrations in the normal state (green) and those in case of failure (red), respectively, for an identical random input was applied to the both (normal and failed) smart beam using a piezoelectric element. The feature extraction of the waveform data was performed by setting sample lines and using ExcelVBA. The MD was calculated using publicly available software. Figure 4 shows the results of fault diagnosis performed using the MT system. The bar graph indicates the MD for each time segment data, and the threshold value is set to 4. It is clearly identified that the system is in failure because the MD exceeds the threshold value many times.

**Results and Discussions**

In this study, fault diagnosis of a smart structure was performed using the random vibration responses and the Mahalanobis-Taguchi system. The experimental results show that the MT system is highly effective in the damage detection.

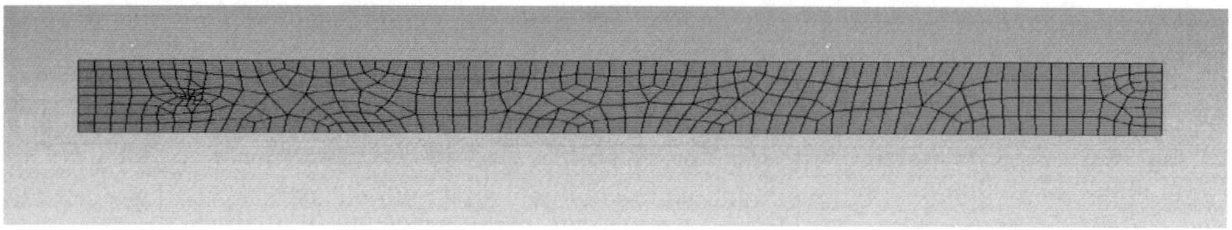

**Figure 1.** Abnormality simulated by an artificial hole near the fixed end of the beam (ANSYS model).

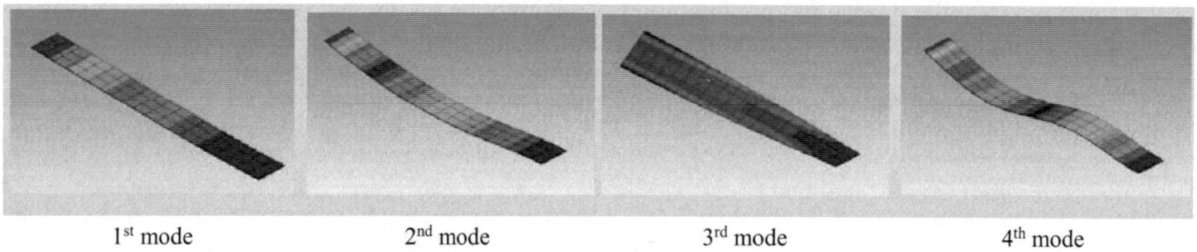


**Figure 2.** Mode shape of the beam.

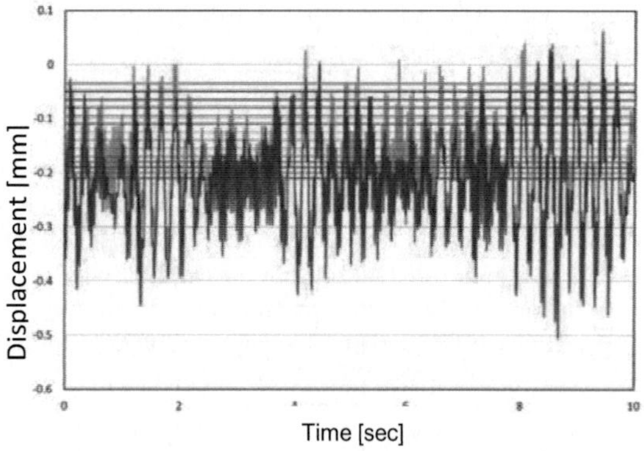


**Figure 3.** Displacement response data and sampling lines-Normal (green) / Failed (red).

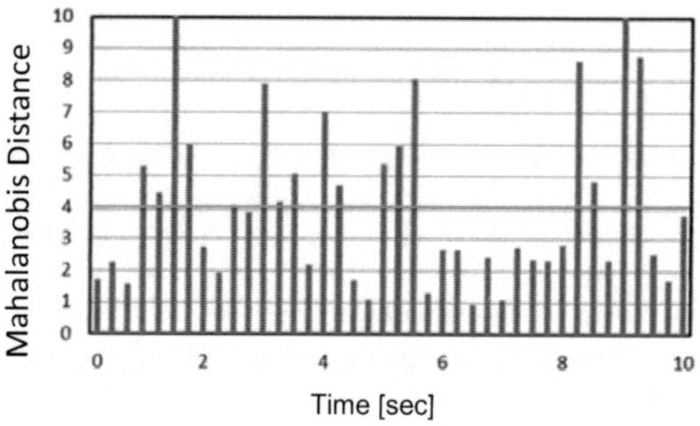


**Figure 4.** MD's for each time segment data (Failed case).

## References

*Taguchi G, Chowdhury S, and Wu Y (2001) The Mahalanobis-Taguchi system. McGraw-Hill Inc.*

*Ushiku T and Okubo H (2018) Application of Taguchi method for fault detection of adaptive structures, 29th International Conference on Adaptive Structures and Technologies, Seoul.*

# Time-reversal Imaging Method based on Laser Ultrasonic Guide Waves with Temporal Filter Method

**Chenguang Xu, Ying Luo*, and Baiqiang Xu**

*Faculty of Civil Engineering and Mechanics, Jiangsu University, National Center for International Research on Structural Health Management of Critical Components, Jiangsu, Zhenjiang, 212013, P. R. China*

**luoying@ujs.edu.cn*

**Abstract**

More attention has been drawn to ultrasonic guided waves (UGW) based damage detection method for its advantages of wide range inspection of large-scale structures. However, complex propagation characteristics of guided waves as well as traditional contact ultrasonic transducers limit its application for the practical damage detection. In this work, A fully non-contact laser ultrasonic detection system is designed by combining YAG pulsed laser with Scanning Laser Doppler Vibrometer (SLDV) to achieve high resolution sensing of UGW field in the structure. A Temporal filter method with multiple frequency bands is proposed to extract the scattering signal without reference signal. The f-k time reversal imaging method is introduced to obtain the reconstructed incident wavefield and scattered wavefield, while the cross-correlation imaging condition is used to identify the damage location in plate structures. Finally, the imaging results of different frequency bands are integrated to achieve accurate imaging with high-precision and high signal-to-noise ratio.

## Introduction

Ultrasonic guided Waves (UGW) based damage detection method is widely used in nondestructive testing and safety assessment of structures because of its advantages of long propagation distance, small attenuation and suitable for wide range detection (Park B, Sohn H and Malinowski P, 2017). However, the contact UGW transducer will significantly affect the mechanical response of the structure, resulting in deviation of the measurement result. In this paper, an all-optical non-contact damage detection platform is constructed by combining laser ultrasonic technology with SLDV. An adaptive temporal filter method is proposed to extract multiple narrow band component in the broadband laser ultrasonic signal respectively. The f-k RTM is introduced to realize accurate damage imaging for each frequency component, which is synthesized to achieve further improvement of the imaging accuracy.

## Methods

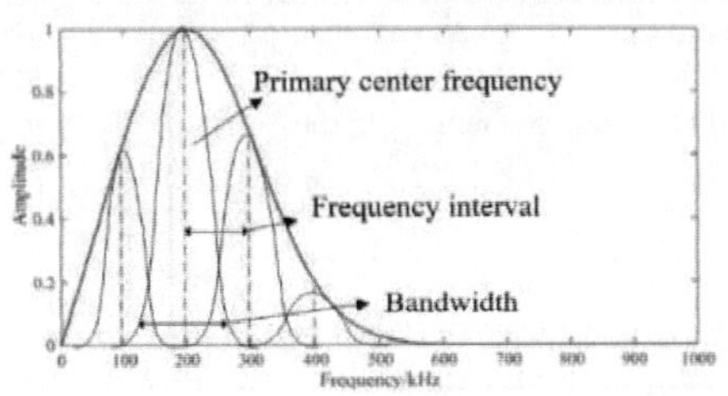


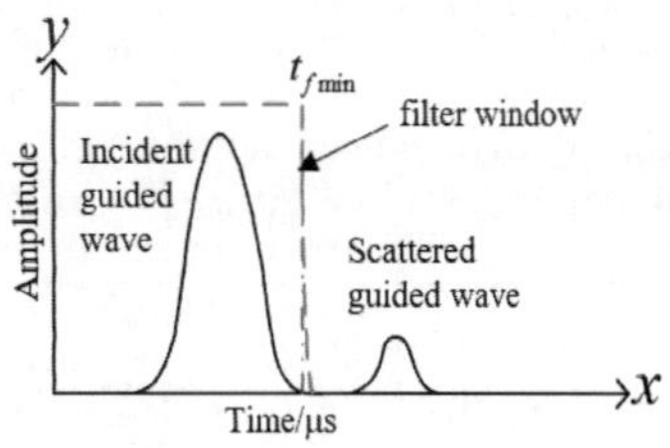


**Figure 1**. Broadband signal decomposition method

The wavelet analysis is used to decomposed the broadband signal into multiple narrowband Signals (Liu Z H, Cao L H and Cun-Fu H E, 2016)

$$WT(a,b) = \frac{1}{\sqrt{a}} \int_{-\infty}^{\infty} S(t)\Psi^* \left(\frac{t-b}{a}\right) dt \tag{1}$$

where $WT(a,b)$ is the separated narrowband component, a and b are scale factor and translation factor respectively, and $\Psi^*(t)$ is the complex conjugate function of the mother wavelet function. Thus, the original broadband signal is decomposed into multiple narrowband component. The width of window function can be determined automatically according to the dispersion relation of the structure and the center frequency of each narrowband signal.

Time reversal imaging method in f-k domain (f-k RTM) reconstruct the scattered wavefield by solving the wave equation in frequency wavenumber domain (Zhu R, Huang G L, Yuan F G, 2013).

$$W^s(x,y,t) = \frac{1}{4\pi^2}\int_{-\infty}^{\infty}\int_{-\infty}^{\infty} W^s(k_x,0,\omega)e^{ik_y y}e^{-i(k_x x-\omega t)}dk_x d\omega \tag{8}$$

As the incident wavefield and the scattered wavefield will be in phase where the reflection happens, the damage can be determined by introducing cross-correlation imaging condition:

$$P_z(x,y) = \sum_t W^i(x,y,t)W^s(x,y,t) \tag{9}$$

where $P_z(x,y)$ is the imaging index of each narrowband signal. The damage imaging results of each narrowband signal are comprehensive analyzed to obtain a more accurate imaging.

## Results and Discussions

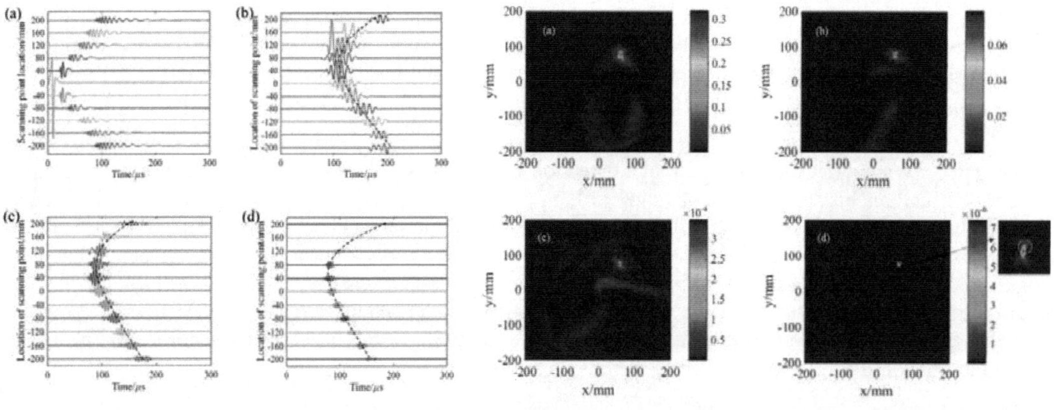


**Figure 2.** Damage imaging results for each narrowband signals (a)100kHz (b)200kHz (c) 300kHz (d) AND operation result

The damage imaging results corresponding to each separated narrowband signal are shown in Figure 2. As the frequency increases, the hotspot area of the imaging results decreases gradually while the noise in the imaging is gradually significant. The signal-to-noise ratio of results by AND operation of each narrowband signals is significantly improved and the partial boundary of the damage is clearly shown.

## References


*Park B, Sohn H, Malinowski P, et al (2017) Delamination localization in wind turbine blades based on adaptive time-of-flight analysis of noncontact laser ultrasonic signals. Nondestructive Testing and Evaluation, 2017, 32(1): 1-20.*

*Liu Z H, Cao L H, Cun-Fu H E, et al (2016) On the Mode Identification and Defect Detection of Laser-induced Lamb Waves Signal in an Aluminum Plate. Journal of Experimental Mechanics.*

*Zhu R, Huang G L, Yuan F G (2013) Fast damage imaging using the time-reversal technique in the frequency–wavenumber domain. Smart Materials & Structures, 22(7):075028.*

# Predictive Analytics of Structural Lateral Distribution Performance for an In-Service Simply Supported Girder Bridge

**Qiwen Jin[1*] and Zheng Liu[2]**

*[1]School of Civil Engineering, Hefei University of Technology, 193 Tunxi Road, Hefei, China.*

*[2]School of Engineering, University of British Columbia, 3333 University Way, Kelowna, Canada.*

*[*]jqw.sinx123@163.com*

## Abstract

So far, Structural health monitoring (SHM) has been widely used in many bridges, with which a lot of SHM data can be collected, such as vibration response and physical response. However, current studies usually focus on structural vibration response. Moreover, further improvement is still needed for the current analysis method, as it is hardly accepted in engineering application to reveal structural performance. This study focuses on the structural performance analysis and prediction of an in-service bridge, which once experienced an explosion caused by a truck loaded with fireworks cannons. The SHM system is installed after several appearance inspections and load tests, and then structural physical response data can be collected. The association analysis method and time series analysis method are considered, with which structural lateral distribution performance can be analyzed and predicted.

## Introduction

The traditional structural performance analysis is usually based on structural vibration response, such as modal response and frequency analysis. However, most of the current codes on structural performance evaluation are based on physical response, such as displacement and microstrain. In this paper, the structural physical response and SHM data mining method are considered (Jin and Bin and Liu, 2018), for the predictive analytics of structural lateral distribution performance of an in-service highway simply supported girder bridge. The bridge was constructed and opened to traffic in 2011. The span and deck arrangement are (5×40) m and (0.5+11+0.5) m, respectively. The accident caused the collapse of south range, and the north range was also damaged. The SHM system was also installed, and a large number of SHM data can then be collected. Taking beam 2, 3, and 5 as the examples, which represent the inside lane, middle lane, and outside lane. Structural lateral distribution performance can be analyzed and predicted.

## Methods

(1) Correlation analysis method

The statistical indicators are usually implemented in correlation analysis, such as correlation diagram and correlation coefficient. The former one is commonly known as the scatter plot, which can be expressed in rectangular coordinate system x and y axis with two variables (Fig. 1). The Pearson correlation coefficient is commonly used in calculation analysis. With the association analysis of SHM data, structural lateral distribution performance of different beams can be reflected in a certain extent.

(2) Holt-Winters time series analysis method

The Holt-Winters method usually needs the current actual value $y_t$ and the single smoothing exponent value $s_t^{(1)}$. Then the prediction of time period $T$ can be realized with a reasonable smoothing coefficient $a$. The prediction model is $\tilde{y}_{t+T} = a_t + b_t \times T + c_t \times T^2$. Where $\tilde{y}_{t+T}$ is the prediction value of period $(t+T)$; $a_t$, $b_t$ $c_t$ are the different model parameters, of which $a_t = 3s_t^{(1)} - 3s_t^{(2)} + s_t^{(3)}$, $b_t = \frac{a}{2(1-a)^2} \times \left[(6-5a)s_t^{(1)} - 2(5-4a)s_t^{(2)} + (4-3a)s_t^{(3)}\right]$, $c_t = \frac{a^2}{2(1-a)^2}(s_t^{(1)} - 2s_t^{(2)} + s_t^{(3)})$. $s_t^{(1)}$, $s_t^{(2)}$, $s_t^{(3)}$ are the single, quadratic, and three smoothing exponent, of which $s_t^{(1)} = a \times y_t + (1-a)s_{t-1}^{(1)}$, $s_t^{(2)} = a \times s_t^{(1)} + (1-a)s_{t-1}^{(2)}$, $s_t^{(3)} = a \times s_t^{(2)} + (1-a)s_{t-1}^{(3)}$.

Additionally, the prediction accuracy can be measured by error analysis, such as Mean Error (ME, $ME=\frac{1}{n}\sum_{i=1}^{n}(Y_i-\hat{Y}_i)$) and Mean Absolute Error (MAE, $MAE=\frac{1}{n}\sum_{i=1}^{n}|Y_i-\hat{Y}_i|$). A smaller prediction error often means a larger accuracy.

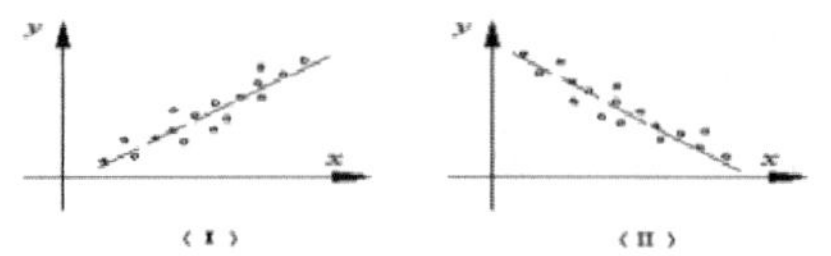


**Figure 1.** Diagram of positive correlation (I) and negative correlation (II).

## Results and Discussions

The diagram of deflection correlation and microstrain correlation of different beams are shown in Figure 2. Where D.b.2, D.b.3, and D.b.5, as well as M.b.2, M.b.3, and M.b.5, are the deflection and microstrain of beam 2, 3, and 5.

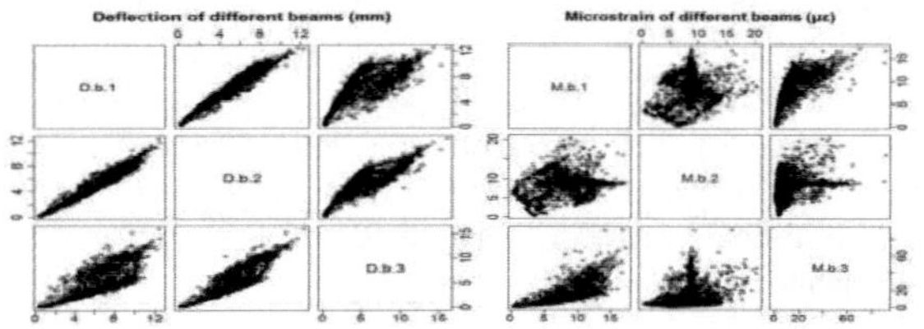


**Figure 2.** The correlation diagram of different beams' deflection (mm) and microstrain.

According to Fig. 2, the deflection correlation of different beams is more significant than the microstrain correlation. It can be speculated that bridge has a good deflection response, or at least, the deflection sensor is more sensitive than the microstrain sensor. Moreover, the correlation of beam 2 and 3 is more significant than that of beam 3 and 5, while the correlation of beam 2 and 5 is poor. It can be speculated that beam 2 and 3 has a good lateral distribution performance, while beam 2 and beam 5 have a weak lateral distribution performance.

Morever, taking the microstrain correlation of beam 3 and 5, from 7 at 0 o'clock to 10 at 24 o' clock as an example. The periodic effect, filtering, and forecast of microstrain correlation is shown in Figure 4. Where the dotted line is the measurement data, the left red line is the method verification. The right blue line and the shaded area is the prediction result with confidence interval. Moreover, the prediction also has a good accuration (ME=0.003, MAE=0.058).

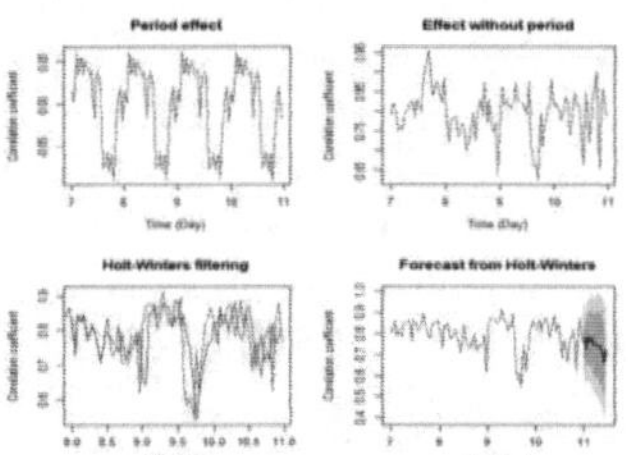


**Figure 3.** The periodic effect, filtering, and forecast of beam 3 and 5's microstrain correlation.

According to Fig. 3, the periodic effect is less than the random effect. The correlation almost has a significant decrement in the afternoon, and it also has a significant increment in the morning. This is due to different traffic volume. The Holt-Winters filtering method can match with the microstrain correlation of measurement data to a certain extent, and then different correlation can be predicted.

## References

*Jin QW, Bin JC, and Liu Z. Structural performance analysis and prediction for in-service bridge with SHM data mining. CSCE 2018 Annual Conference, 1-11, Canada, 2018.*

**Gap in pagination due to unavailable papers.**

**Pages 115-116**

# Estimation of the Capacity of a Bridge Deck and the Bridge Superstructure System based on Ground Penetrating Radar Imaging

**Dipesh Donda*, Farzad Ghodoosi, and Ashutosh Bagchi**

*Building, Civil and Environmental Engineering Department, Concordia University, Montreal, Quebec, Canada- H3G 2W1*

**dipeshdonda@gmail.com*

**Abstract**

The load carrying capacity of the bridge reduces with time, both in shear and flexure. Such capacity reduction is due to the deterioration of concrete and steel reinforcement in reinforced concrete bridges. Corrosion of steel reinforcement results in cracks, concrete delamination, and spall. Identifying the deterioration pattern of a bridge deck helps decision-makers to come up with cost-effective repair schedules. The existing Bridge Management Systems (BMSs) such as BRIDGIT, POINTIS, and MTQ rating system are based on subjective visual inspections. The visual inspection of reinforced concrete bridges often results in an inaccurate condition assessment. Moreover, the traditional approach in Bridge Management Systems includes the deterioration of a bridge at the element level. Due to the redundancy and load redistribution, the actual system capacity is larger than the element capacity. Due to the development of the Non-Destructive Testing (NDT) methods, the actual deteriorated condition of the bridge could be assessed more accurately. The objective of this research is to analyze the capacity of the bridge using the Ground Penetration Radar (GPR) image. Here, the deck GPR map is used to analyze the actual flexural and shear capacity of the bridge deck, and the whole superstructure as a system. The GPR map is further used to simulate the possible deterioration patterns, to analyze the components of the bridge and to estimate the capacity of the entire bridge system. Here, the comparison is to be made between the capacity of the bridge deck elements and the superstructure system for different types of deterioration pattern. The proposed method will be useful for determining the capacity of a bridge for different time intervals by using the deterioration prediction curves. This approach will be useful in determining the degree of repairs needed for the deck elements from the deterioration model with time.

## Introduction

Bridges are important for the economic growth of any country. The bridges get deteriorated with time and thus its structural health monitoring is essential. The existing Bridge Management Systems are based on visual inspection which concludes to subjective results. The use of Non-Destructive Evaluation methods results in the more precise condition assessment of the bridge. The objective of this study is to adopt the GPR maps to simulate the deterioration patterns in beams and slab for the purpose of objective structural analysis. Cracks, top and bottom rebar corrosion in beams and slab and concrete spall are detected based on the GPR test; consequently, those findings are applied to a non-linear finite element model to analyze the superstructure. (Ghodoosi et al, 2018)

## Methodology

In this study, the GPR defect maps similar to GPR map shown in Fig. 1 are used to assess the condition of a bridge in terms of flexural capacity and shear capacity. Different GPR maps are produced based on computer simulations. These GPR maps are adapted to measure the load carrying capacity of bridge elements and the system by performing the pushdown non-linear analysis. Location of shear and flexural hinges for calculation of capacity is considered similar to Fig. 2.

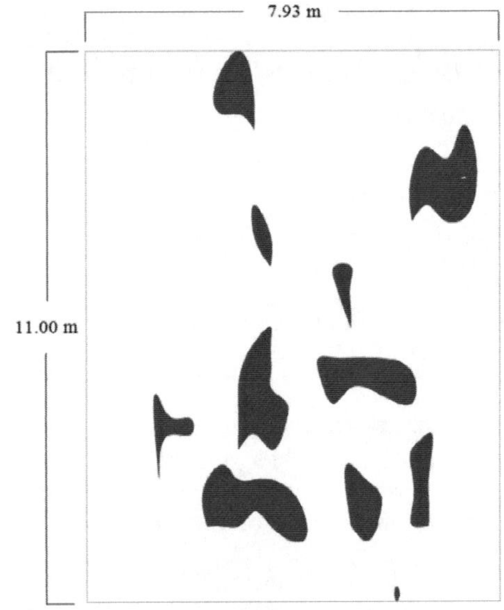


**Figure 1**. Map based on GPR results. (Ghodoosi et al, 2018)

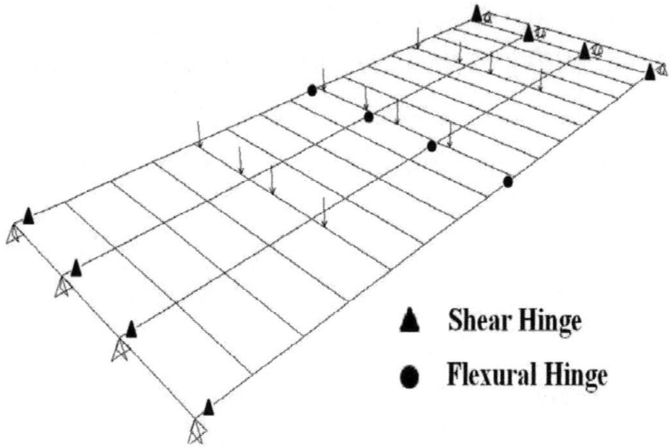


**Figure 2**. Finite Element model using Grillage analogy, two side-by-side trucks are loaded on the bridge

## Results

This study develops a method to implement the results of a GPR test into a finite element structural analysis model to measure the capacity of a bridge superstructure in an objective manner. The results will help decision-makers to plan for a bridge repair and rehabilitation while the whole structural system is analyzed based on element conditions obtained through a Non-Destructive GPR test result.

## References

*Ghodoosi, F., Bagchi, A., Zayed, T. and Hosseini, M.R., 2018. Method for developing and updating deterioration models for concrete bridge decks using GPR data. Automation in Construction, 91, pp.133-141.*

30th International Conference on
Adaptive Structures and Technologies

# Automated Visual Assessment of Structural Conditions by FE Model Updating using Building Information Modeling (BIM)

**S. Bahmanoo*, M. Valinejadshoubi, F. A. Sakib, A. Sabamehr, A. Bagchi, and A. Bhowmick**

*Dept. of Building, Civil and Environmental Engineering, Concordia Univ., 1455 de Maisonneuve Blvd. West, Montreal, QC, Canada*

**sam_bahman@yahoo.com*

**Abstract**

Vibration-Based Damage Identification (VBDI) is an important class of techniques used for Structural Health Monitoring (SHM) of aerospace, civil and mechanical engineering structures. According to VBDI premise, the deviation in structural properties such as mass, stiffness and damping ratio could be associated by adverse changes on global vibrational response of structure as an indication of damage. Thus, finite element model updating technique has been widely adopted as a part of VBDI process to obtain and evaluate up-to-date structure model, structural properties and performance. However, using a visual data management system to collect and categorize the updated structural properties as well as the corresponding global vibration signatures such as natural frequencies and mode shapes in subsequent states is yet to be fully realized. Therefore, utilizing Building Information Modeling (BIM) as an integrated data management system would be a great step towards enhancing the efficiency of finite model updating technique in terms of data presentation and decision making. In this study a vibration-based SHM framework is demonstrated through adopting a FE model of a 4-story steel frame as the case study.

**Introduction**

Few studies have been conducted on establishing an automated visual data management system named Building Information Modeling (BIM) in Structural Health Monitoring (SHM) implementation. Among those studies, although BIM-based approaches are well adopted and established for strain-based structural monitoring, its full potential in vibration-based SHM has not been extensively studied yet (Yang and Bai 2015). This paper is a state-of-the-art-study, proposing a BIM-based SHM system that visualizes the derived structural properties from experimental modal analysis and FE model updating process to boost structural condition assessment in buildings. Here, a 4-story steel frame is used for demonstrating how BIM could be used for managing SHM data in a building and render the results of damage localization. Initially the frame is modeled in SAP2000 to obtain modal properties such as natural frequencies and the corresponding mode shapes. Then by applying White Gaussian Noise on the FE model and performing experimental modal analysis in MATLAB, measured modal properties are derived. At the next step, by adjusting the derived frequencies between the initial and the measured states through structural parameters modification using M-FEM (a MATLAB based FE program), stiffness adjustment factors for all structural elements are derived. Eventually, Dynamo Studio, a visual programming tool is utilized to assign automatically the derived stiffness adjustment factors to the structural elements in Autodesk Revit as a BIM software based on a predefined color-coding scheme. As the result, an automated BIM-based linkage would be created that could visualize the initial and updated FE model characteristics for better comparison of the changed data in structural parameters at subsequent states.

**Methodology**

Figure 1 is demonstrating the methodology of vibration-based SHM and BIM integration proposed in this study.

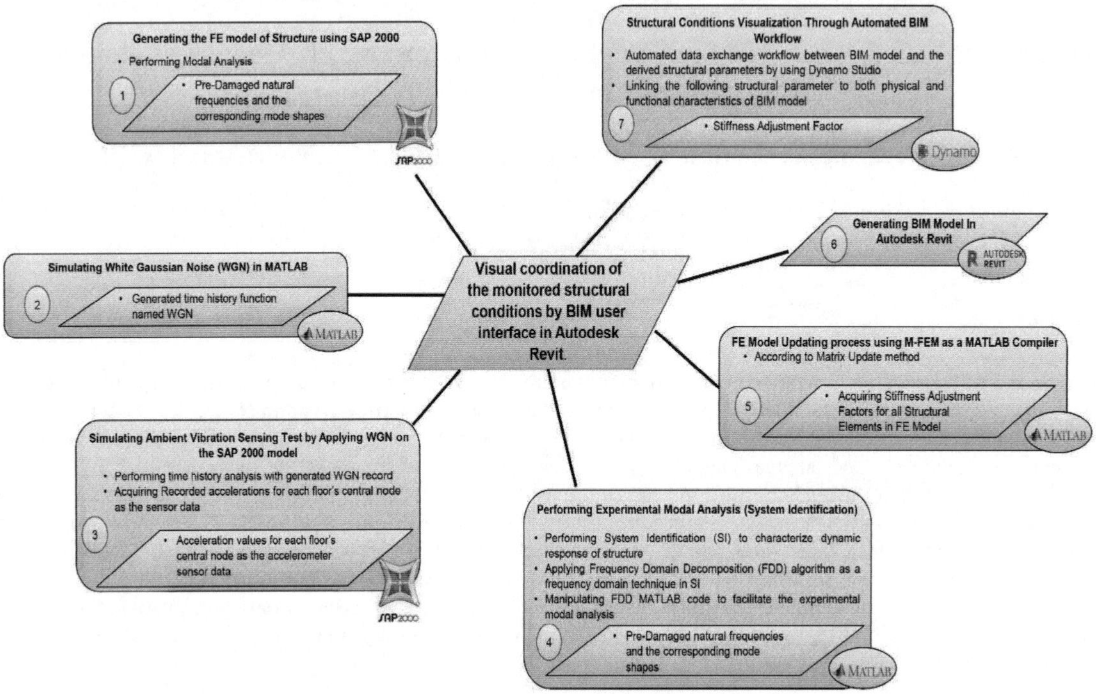


**Figure 1**. Proposed methodology to visualize vibration-based SHM process in BIM environment.

## Results and Discussions

This paper proposed a visual automated BIM-based framework in vibration-based SHM that boost remarkably structural condition assessment in civil engineering structures in terms of data presentation and decision making. With regards to Fe model updating process as a fundamental principle in vibration-based SHM, visualizing the baseline FEM model has been always a challenge. However, the developed structural assessment system in this study is a BIM-based user interface that could visualize and assign the derived structural parameters such as stiffness adjustment factor to the corresponding structural elements to ease structural condition perception particularly in the presence of probable damages.

## References

*Zhang, Y. and Bai, L. 2015, "Rapid Structural Condition Assessment Using Radio Frequency Identification (RFID) Based Wireless Strain Sensor", Automation in Construction, vol. 54, pp. 1-11.*

# SHM Point Arrangement for In-service Continuous Bridge with Consideration of Structural Robustness

**Qiwen Jin[1*] and Zheng Liu[2]**

[1]*School of Civil Engineering, Hefei University of Technology, 193 Tunxi Road, Hefei, China.*

[2]*School of Engineering, University of British Columbia, 3333 University Way, Kelowna, BC, Canada.*

[*]*jqw.sinx123@163.com*

**Abstract**

In this paper, a strain energy evaluation index of structural robustness is introduced firstly. A numerical simulation for structural robustness of an in-service bridge subjected to different moving speeds is then performed. Different damage location and damage degree are considered. Then, the elements with a smaller structural robustness are located as the vulnerable part, and can be selected as the SHM monitoring point.

## Introduction

The SHM monitoring point arrangement is usually based on traditional load tests and damage analysis. Hundreds of monitoring sensors are usually employed, which will not only need a high engineering cost but also increase the follow-up maintenance task of the SHM system. The mass monitoring data will also be difficult to process effectively. All these negative questions will affect the application effect of SHM and make it difficult to be more widely used. Some new studies gradually concentrate on structural robustness analysis. The vulnerable and weak parts of a bridge structure can then be located to arrange the different SHM monitoring sensors [1].

## Strain Energy Evaluation Index for Structural Robustness

According to the universal applicability of energy principle and the different expression forms of current evaluation indexes of structural robustness, the expression of structural robustness (R) can be shown as follow [2]:

$$R = 1 - \frac{U}{U_M} \tag{1}$$

where $U=\int_0^l \frac{M(x)^2}{2EI}dx$ is the bending strain energy of a single beam generated by the bending moment $M(x)$, and $U_M=\int_0^l \frac{M_{limit}^2}{2EI}dx$ is the limit strain energy of the beam. $EI$ represents the structural stiffness, and $l$ is the length. $M_{limit}=\frac{2\sigma I}{h}=\frac{\sigma bh^2}{6}$ ($\sigma$ is the material yield strength. $h$ and $b$ are the height and width of the cross section, respectively).

## Finite Element Numerical Simulation

We take a continuous girder bridge with two spans (7×2 m) and three spans (7×3 m) as the example. Fourteen elements and twenty one elements are discreted, respectively. The ANSYS finite element of the continuous girder bridge with one bridge span (the simply supported girder bridge, ten elements are discreted) is shown in Fig. 1 [2]. Two kinds of loss degrees of elastic modulus are considered, of which I=5% and II=15%. With the consideration of the maximum designed traffic speed of the highway bridge in China (120 km/h, i.e. 33.3 m/s), five different moving speeds are considered. V1=5 m/s, V2=10 m/s, V3=20 m/s, V4=30 m/s and V5=35 m/s. The moving vehicle load is represented by the concentrated force (4 kN). In this section, the transient dynamic finite element analysis is considered. A total of twelve working conditions are considered, including the initial good condition and the damage effect at different speeds (Table. 1).

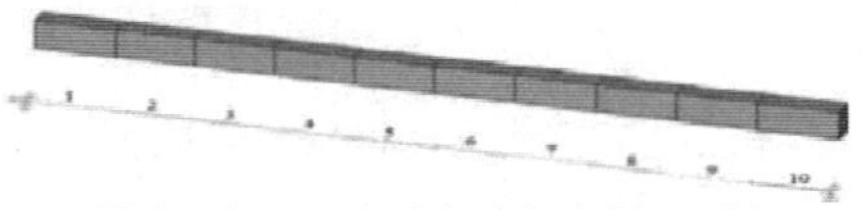

**Figure 1.** Finite element model of the bridge with one span

**Table. 1** Different working conditions of structural robustness

| Working condition | | Damage element | Damage degree | Moving speed |
|---|---|---|---|---|
| Two spans | 1 | / | Initial good condition | From V1 to V5 |
| | 2 | From *E*.1 to *E*.14 | I, II | V1 |
| | 3 | From *E*.1 to *E*.14 | I, II | V2 |
| | 4 | From *E*.1 to *E*.14 | I, II | V3 |
| | 5 | From *E*.1 to *E*.14 | I, II | V4 |
| | 6 | From *E*.1 to *E*.14 | I, II | V5 |
| Three spans | 7 | / | Initial good condition | From V1 to V5 |
| | 8 | From *E*.1 to *E*.21 | I, II | V1 |
| | 9 | From *E*.1 to *E*.21 | I, II | V2 |
| | 10 | From *E*.1 to *E*.21 | I, II | V3 |
| | 11 | From *E*.1 to *E*.21 | I, II | V4 |
| | 12 | From *E*.1 to *E*.21 | I, II | V5 |

## Results and Discussions

For the bridge with one span, the structural robustness of different elements at different moving speeds under initial good condition is shown in Fig. 2 [2]. Where the plotted values for each element correspond to the value corresponding to the final time, when the load is applied to the element. Different from the simple bending moment diagram, the structural robustness is the change rate of strain energy (the strain energy is also calculated by the bending moment). The weak part of the beam can then be reflected by the smaller structural robustness. Even the moment value can meet the service requirements, more attention should still be paid on the weak part where the change rate of the strain energy is larger. Moreover, a basic comparison can also be provided for the following analysis of the bridge span subjected to different damage effect.

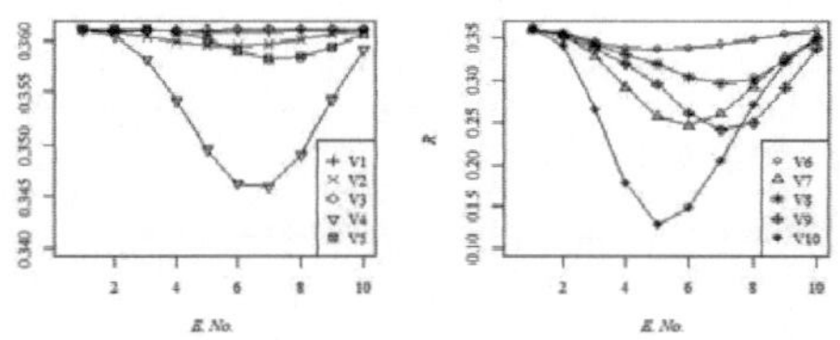

**Figure 2.** Structural robustness of different elements at different speeds under initial good condition

According to Fig. 2, the structural robustness of different elements at different speeds shows a significant change with different variation trends, except for speed V1 (1.25 m/s) and V3 (5 m/s). The smallest structural robustness of different elements at different speeds is almost located around the mid-span region. The corresponding velocity of the smallest structural robustness is V10 (35 m/s). Therefore, the weak part is around the mid-span region. More attention should be paid to the mid-span region, especially at speed 35 m/s.

## References


*1. Yu G. Structural vulnerability analysis and its application in bridge health monitoring[J]. Shanghai, China: Tongji University, 2009.*

*2. Jin Q.W. and Liu Z. In-service bridge SHM point arrangement with consideration of structural Robustness[J]. Journal of Civil Structural Health Monitoring, 2019 (Published Online).*

# Experimental Broadband Signal Reconstruction for Plate-like Structures

**Nicolas Venturini[1,2]**, Marcias Martinez[1,3]*, Enrico Troiani[2], Maria Barroso-Romero[3]***, and Francesco Falcetelli[2]**

*[1] Clarkson University, Department of Mechanical and Aeronautical Engineering, Potsdam, NY, 13699, USA.*

**mmartine@clarkson.edu*

*[2] University of Bologna, Department of Industrial Engineering, via Zamboni 33, 40126, Bologna, Italy.*

***nicolas.venturini@studio.unibo.it*

*[3]Faculty of Aerospace Engineering, Delft University of Technology, Kluyverweg 1, 2629 HS Delft, Netherlands*

****M.BarrosoRomero@tudelft.nl*

**Abstract**

In the Structural Health Monitoring (SHM) field, Acoustic Emissions (AE) is the process by which acoustic signals generated during the formation of damage are captured by sensors, analyzed and used for localization within the structure. In plate like structures, these signals lead to the formation of Lamb Waves (LW), which are broadband in nature. These LW are generally captured by Piezoelectric Titanum Zirconate (PZT) sensors. As such, the captured broadband signals are of difficult interpretation in part due to several phenomena such as dispersion or attenuation suffered by the waves during their propagation. In this study, we hypothesize that the nature of the emitted signal contains information on the damage type, as if the features of the emitted signal were a 'fingerprint' of the damage. Wing or fuselage panels are some of the aeronautical structures were LW can develop during the emission of an acoustic signal. In operational service environments, the damage type and size may lead to the generation of different signal sources. This study aims at the development, through experimental techniques, of a classification algorithm based on Artificial Intelligence (AI) for determining the source of the emission in addition to their location within a structure. It is envisioned that the AI algorithms will be capable of identifying specific features within the emitted signals and thus correlate them to a database of known signals and their corresponding associated damage types. In order to create an AE signal damage database, the captured signal cannot be used since it has been affected by its propagation through the structure. As such, a Time Reversal process will be implemented in order to reconstruct the original signal. This original signal will be the one utilized by the AI algorithm in order to identify its corresponding damage source.

In order to achieve this objective, flat plates used as a simplification of aerospace structures were coupled with a PZT sensor network connected to a digital multi-channel AE Vallen Système. The piezoelectric sensors and actuators were provided by PiezoKinetics Inc. and used to form the sensor network. These sensors are connected to the AE system through pre-amplifiers. A series of tests signals with a corresponding central frequency and broadband in nature were used during our experiments. These signals consisted of Hanning Windows (HW) and Pencil Lead Breaks (PLB). The HW signals consist of a central frequency from 50kHz to 500kHz. The HW signals are generated, amplified and emitted thanks to a digital Picoscope 3204-DMSO (oscilloscope), the ENI 240L Broadband Power Amplifier and the PZT actuator respectively as shown in Figure 1.a, 1.b and 1.c. The PLB signals were performed following American National standard ASTM 976. The AE generated where captured by the PZT sensors which formed the sensor network as shown in Figure 1.d. Furthermore, to reduce the signal reflection phenomenon and to record a clear signal, the plates were bordered by plasticine as presented in Figure 1.c.

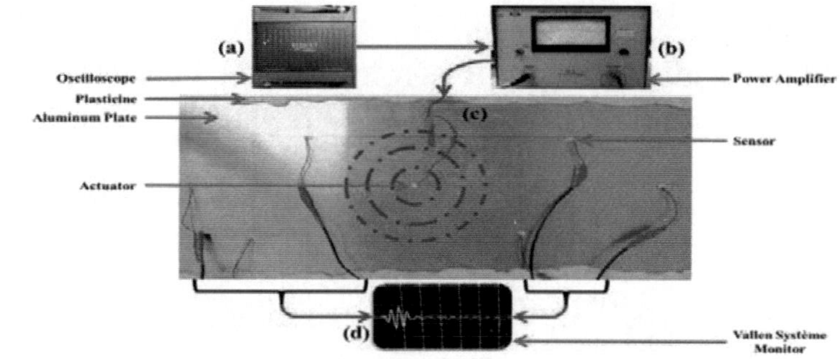


**Figure 1.** a) The signal is sent from the oscilloscope to the power amplifier - b) the signal is amplified by the power amplifier - c) the amplified signal is emitted by the PZT actuator and the LW propagate in the plate surrounded by plasticine - d) the PZT sensors received the signals which were then captured by the AE Vallen Systéme.

The data processing part of the received signal was performed in MATLAB$^{TM}$. The mode identification is based on the Hilbert transform of the received signals. Firstly, the envelope of the received signal is generated. This allows the identification of the time of arrival (ToA) of the $S_0$ mode. Secondly, the phase and the instantaneous frequency variations of the signal are computed. Slight variation of the slope of the phase curve and a discontinuity in the instantaneous frequency of the signal allows for identification of the $A_0$ wave packet. From the identification of the $S_0$ and $A_0$, it was possible to create a repeatable time window which would be time reversed.

Moreover, thanks to the knowledge of the ToAs, a traditional triangulation algorithm as discussed in the literature was implemented in order to determine the origin of the emission. Finally, the time reversed signal is amplified and re-transmitted to the origin of the source with a set-up analogous to the one presented in Figure 1. For signals broadband in nature, appropriate transfer functions related to the experimental set-up was needed in order to account for augmentation and reduction contributions to the signal amplitude at different frequencies. The validation of the results is performed through a cross-correlation of the reconstructed signal with the original one for both the HW and the broadband signal. Thanks to a comparison of signals, it is possible to confirm the reconstruction of the broadband signal as shown in Figure 2 where the $S_0$ wave packet of a real PLB signal received in one sensor is compared with the $S_0$ wave packet of a reconstructed PLB signal sent from an actuator and received by the same sensor.

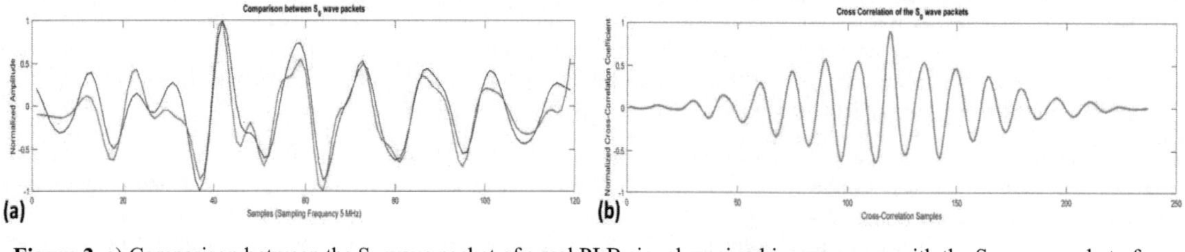


**Figure 2.** a) Comparison between the $S_0$ wave packet of a real PLB signal received in one sensor with the $S_0$ wave packet of a reconstructed PLB signal sent from an actuator and received by the same sensor.- b) Normalized cross-correlation between the two $S_0$ wave packets reaching a maximum of more than 0.90.

# Vibration-based Damage Detection in a Cable-stayed Bridge

**Zahrasadat Momeni* and Ashutosh Bagchi****

*Department of Building, Civil & Environmental Engineering, Concordia University*

*Montréal, QC, H3G 1M8, Canada*

**zahra.momeni24@yahoo.com* , ***ashutosh.bagchi@concordia.ca*

## Introduction

Cable-supported bridges are characterized by a long fundamental period; large flexibility and low structural damping; a combination which makes this type of bridges vulnerable in large amplitude. Vibration monitoring and modal analysis play an important role in identifying structural conditions of such structures and aid in detection of damage or distress. Modal analysis involves the field measurement of vibration response, and estimation of natural frequency and mode shape vectors through a process of system identification. In this study, one of the mode shape based technique which is called Mode Shape Curvature (MSC) method has been used to detect the damage [Bagchi et al 2016].

## Methodology

In this study, the Quincy Bayview Bridge on the Mississippi river in Illinois in US was selected and has been modeled in ABAQUS software. The ABAQUS software is used for 3D-modeling and Frequency analysis of the considered cable-stayed bridge. The artificial earthquake records is considered taking into account the distance between the piers of the considered bridge in two directions. Two damages were made by changing the properties of the elements of the structure and the performance of three damage detection methods to detect the considered damage including Natural Frequency-based Methods and Methods based on mode shape changes and Method based on mode shape curvature were investigated.

## Results

### Natural Frequency-based Methods

Natural frequency-based methods is based on natural frequency change for damage identification. Table 1. shows frequencies of undamaged and damaged bridge. It is obvious all frequencies are reduced after damage, but the differences are very subtle and location of damage cannot be identified.

### Methods based on Mode Shape Changes

In this study, the mode shapes of structure are extracted and plotted in excel. As it is obvious in Fig.1 and Fig.2 the mode shapes of intact and damaged structure are as same as each other.

### Methods based on Mode Shape Curvature/ Strain Mode Shape Changes

Mode shape curvature is an alternative approach to determine vibration changes that is based on the curvature. The first and third mode shape curvature are prepared in Fig. 3 and Fig.4.

**Table 1**.Frequencies of undamaged and damaged bridge

| Mode no. | undamaged (Hz) | damaged (Hz) | Difference (%) |
|---|---|---|---|
| 1st | 0.36484 | 0.36468 | 0.04 |
| 2nd | 0.39595 | 0.39501 | 0.02 |
| 3th | 0.41751 | 0.41731 | 0.04 |
| 4th | 0.50402 | 0.50400 | 0.004 |
| 5th | 0.59742 | 0.59563 | 0.2 |
| 6th | 0.62145 | 0.62143 | 0.03 |
| 7th | 0.66457 | 0.66455 | 0.03 |
| 8th | 0.71295 | 0.71014 | 0.3 |

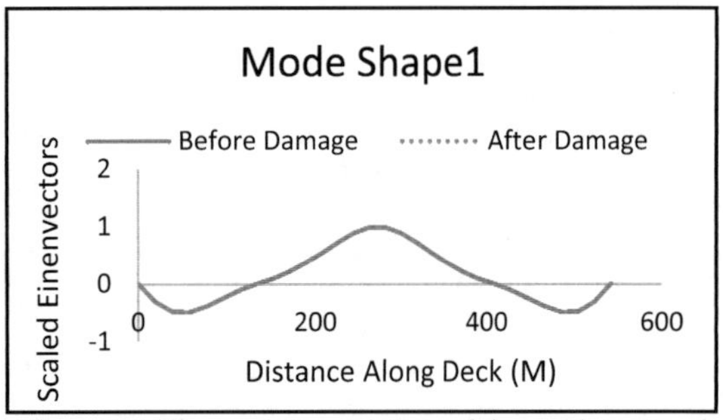


**Figure 1**. Difference between intact and damaged structure in first mode shape

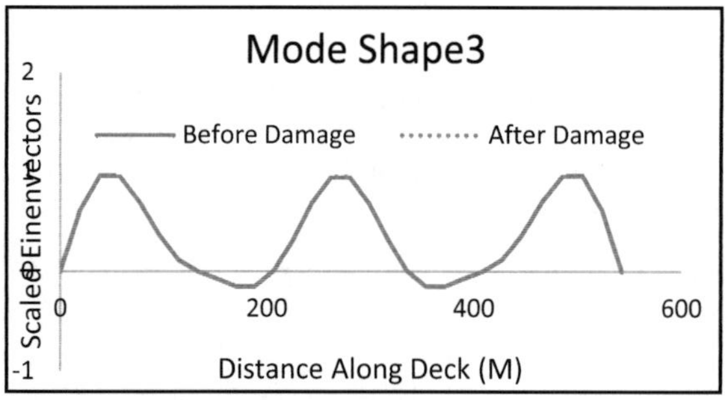


**Figure 2**.Difference between intact and damaged structure in third mode shape

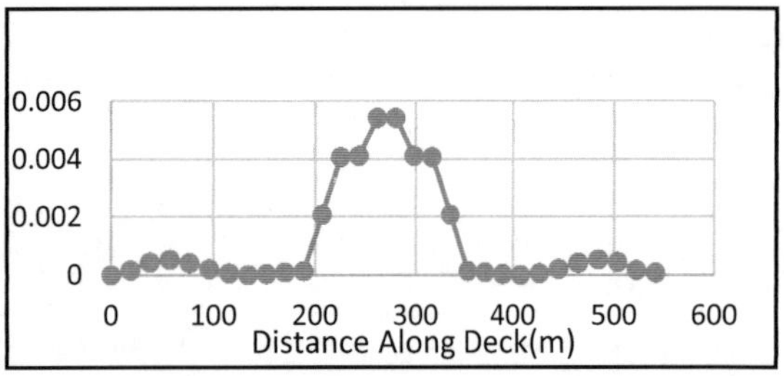


**Figure 3**. Difference between intact and damaged structure in first mode shape curvature

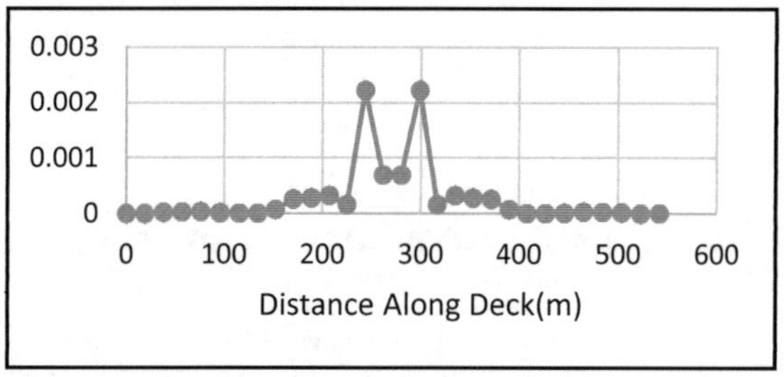


**Figure 4**.Difference between intact and damaged structure in third mode shape curvature

This study investigate the effectiveness of some damage identification methods such as natural frequency-based method, methods based on mode shape changes and methods based on mode shape curvature. Natural frequency-based method is not reliable for detecting the damage in complex and huge structure such as cable stayed bridge. The differences in frequencies is subtle and location of damage cannot be identified. Mode shape change method is in practical in subtle damages. Mode shape curvature method, however, is effective to find out the location of damage.

## Reference

*Ch. Lim, A. Sabamehr, A. Bagchi(2016), "SYSTEM IDENTIFICATION AND DAMAGE DETECTION TEQNICUES IN PRESTRESS CONCRETE BOX BRIDGE",CSCE*

# System Identification of Concordia University EV Building by Operational Modal Analysis

**Saikat Bagchi*, Ardalan Sabamehr, Timir Baran Roy, and Ashutosh Bagchi**

*Department of Building, Civil and Environmental Engineering, Concordia University, Montreal, Quebec, Canada*

**er.saikat.ac@gmail.com*

**Abstract**

Identification of the behavior of structural system and the monitoring of its structural health is essential for any smart and sustainable city. Montreal being one of the premier urban destinations situated at one of the extreme locations on earth, calls for such facilities to have installed for its important structural systems to obtain real-time information about the state of functionality of the structure. In this study, the EV Building of Concordia University (SGW Campus, Montreal, Canada) is studied using Operational Modal Analysis to comprehend the behavior of its two sub-systems. Frequency Domain Decomposition (FDD) technique is employed to extract the modal information.

**Introduction**

System Identification (SI) and monitoring of structural health of a civil engineering structure have emerged as an essential engineering requirement for any smart and sustainable city. Urban facilities of Montreal is known to be one of the advanced most among the world. On the other hand, the city is the home for the age old civil infrastructures as well. Naturally, with the advent of Civionics there is a constant demand for smart structural health monitoring system, which can help the structural engineers, building owners and city authorities to comprehend the properties, behavior and performance of these diverse infrastructures.

One of the most practiced methods of SI is Operational Modal Analysis (OMA), where ambient vibration response of the structure is recorded and analyzed to comprehend the state of its functionality. As in case of OMA information about the input excitation is not available, choice of appropriate signal processing methodology is essential to extract information pertaining to the structure. The classical technique of signal processing, which is based on Direct Fourier Transform, can clearly identify the modes from the response signal only if they are separated enough from each other (Brincker et al., 2000). Additionally, estimation of modal damping is either quite uncertain or impossible in this method. The classical method, however, is used quite frequently because of its user-friendliness in comparison to some other techniques like Ibrahim Time Domain Technique, Stochastic Subspace Method, etc. Frequency Domain Decomposition (FDD), on the other hand, is such a methodology which uses Singular Value Decomposition (SVD) of the spectral matrix to extract the modal information.

The current article reports the study conducted on the EV Building of Concordia University (SGW Campus), Montreal, Canada. Fig. 1 (a) shows the location of EV building with its two partially interconnected units such as, faculty of 'Gina Cody School of Engineering and Computer Science' (ENCS) and faculty of 'Arts'. FDD based SI technique is employed in this study to identify behavior of the coupled building systems.

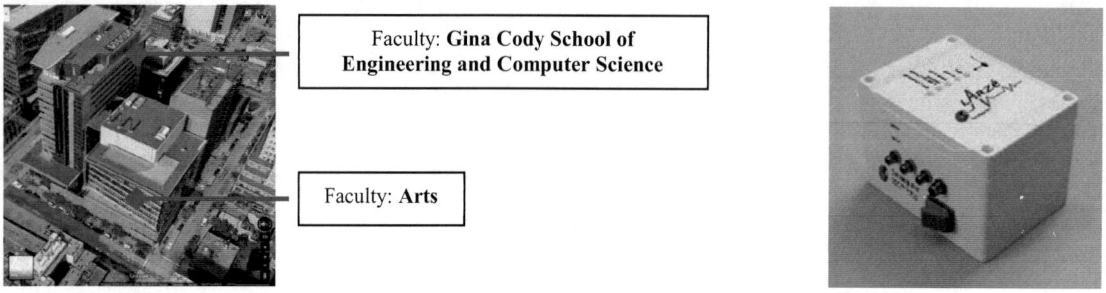


(a) (b)

**Figure 1.** (a) Isometric view of the EV Building, Concordia University (Source: Google Map), (b) Larze sensor (Copyright: Sensequake Inc.)

### Methodology

Response of the building units are measured using 'Large Sensor' made by 'Sensequake Inc. From 2nd floor onwards each alternate floor (i.e., 2nd, 4th, 6th and so on) up to 10th floor was equipped with three sensors simultaneously. Among them two were placed at the ENCS unit and the other one was at the Arts unit. As evident from the Fig. 1, ENCS unit is taller than the Arts unit, which is up to the 11th storey. Therefore, 12th Storey onwards only two sensors, i.e., in the ENCS part, were used. As response of the structure is recorded against the ambient vibration, output only SI technique FDD is employed to extract the modal information. Matlab based FDD code is utilized for this purpose.

### Results and Discussions

Modal response of the structure are extracted for the combined units using the responses from all three sensors (Table 1) as well as for each of the two units apiece. It is intended to identify the structural integrity between the two units. Preliminary analysis on the results show that the units which are believed to be disjointed are responding together. Therefore, the connection between them which appears to be of non-structural type may have sufficient stiffness to connect the units together while participating in the modal vibration.

**Table 1.**

| Mode | Natural frequency (Hz) | Description |
|---|---|---|
| First | 0.66 | Lateral mode in 'X' and 'Y' direction |
| Second | 0.73 | Lateral mode in 'X' and 'Y' direction |
| Third | 0.84 | Torsional Mode |

### References

*Brincker, R., Zhang, L., and Andersen, P. (2000, February). Modal identification from ambient responses using frequency domain decomposition. In Proc. of the 18th International Modal Analysis Conference (IMAC), San Antonio, Texas.*

# Shear Behaviour of Magnetorheological Elastomers: Viscoelastic and Magnetorheological Properties

**Andrea Spaggiari[1,2*] and Alberto Bellelli[2]**

[1]*Centro Interdipartimentale Intermech More – University of Modena and Reggio Emilia*

[2]*Dipartimento di Scienze e Metodi dell'Ingegneria – University of Modena and Reggio Emilia*

[*]*andrea.spaggiari@unimore.it*

## Introduction

The magnetorheological elastomers (MRE) are a class of smart materials which modifies their mechanical properties due to an external stimulus, the magnetic field. These materials are gaining increasing popularity nowadays, since the possibility to have free shape MREs could lead to the design of smart components or sensors, which exhibits full integration within the structure. Therefore, the proper design of an MRE system first requires the knowledge of the magneto-mechanical properties of the MRE material. These materials have already gained the attention of the researcher, but still some open questions remain in technical literature about their magneto-mechanical properties. Magnetorheological elastomers (MREs) consist of a non-magnetic elastomeric matrix with ferromagnetic micrometric particles. The technical literature reports several studies in compressive behaviour, considering both dynamic and quasi static conditions, but only scarce information are retrieved in shear (Schubert and Harrison, 2015) and especially regarding the ultimate failure stress, since the shear test requires an ad hoc design of the gripper system and the DMA technique cannot be used. The purpose of this work is to expand the applicability of MREs in the industrial world, as already happens for the magnetorheological fluids (MRFs), with the benefit of the solid behaviour of the MRE, since they do not switch from fluid to quasi-solid state, however their mechanical characteristics in terms of stiffness and damping are a function not only of the external mechanical loads, but also of the applied external magnetic field. In order to fully exploit MRE capabilities, this work investigated the static and dynamic properties of MREs in shear mode, one of the most efficient way to exploit an elastomeric structure. Typically, the technical literature available is focused on the properties of these system in the small strains, but this paper provides information about the characterization up to failure of these materials, both under monotonic and in cyclic loading.

## Materials and Method

The manufacture of the specimens was carried out following a Design of Experiment procedure. We considered as design variables three different volume fractions of ferromagnetic particles, in addition to a pure base material one, the isotropicity of the specimens and the application of an external magnetic field, which is provided with permanent magnets. For each configuration we manufactured 2 samples, for a total of 14 specimens. Subsequently, the shear tests, with or without an external magnetic field, are carried out both in monotonic and cyclic loading up to failure.

The specimens were tested in shear, by means of a classical pin and fork rig, as shown in Figure 1. This configuration allows a quasi-pure shear stress on the MRE without any need of clamps or adhesives. The complete system is reported in Figure 1b. The shear stress test procedure was divided in two parts. The first set of tests were non-destructive since we decided to limit the test to the elastic properties of the material and no plasticization is expected for a hyper-elastic material. The desired test conditions were ensured by limiting the maximum force to 500 N and applying a cyclic displacement.

## Results and Discussion

The results of the first test are reported in Figure 2. Figure 2a show the cyclic test on the MRE at low speed in terms of shear stress, $\tau$, versus shear strain, $\gamma$. This two quantities are computed as follows, where $d$ is the diameter of the MRE pin, $h$ is the gap between the fork and the rod, $u$ is the applied displacement and $F$ is the tensile force.

$$\tau = \frac{2F}{\pi d^2} \qquad\qquad \gamma = \frac{u}{h}$$

The preliminary results show that the proposed rig is able to apply both cyclic and monotonic loading to the MRE specimen in shear mode without bonding or other constraints which modifies the MRE behaviour due to the incompressibility of the matrix.

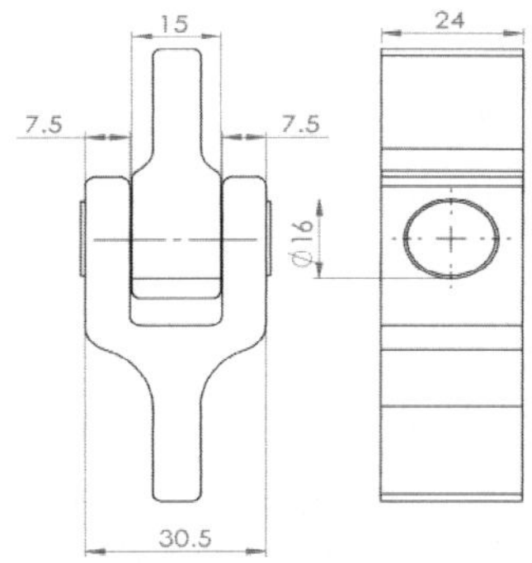


(a)

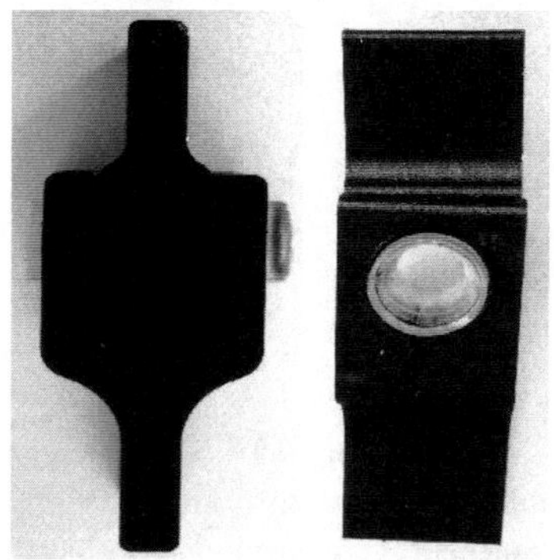

(b)

**Figure 1.** Schematic of the test rig with dimensions (not to scale) and 3D printed test rig with pure elastomeric specimen.

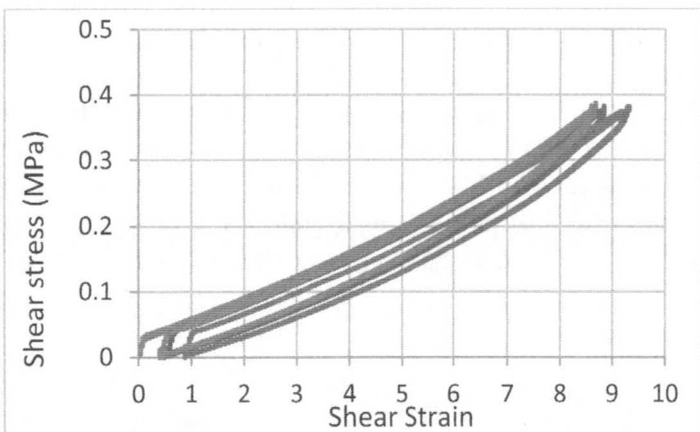


(a)

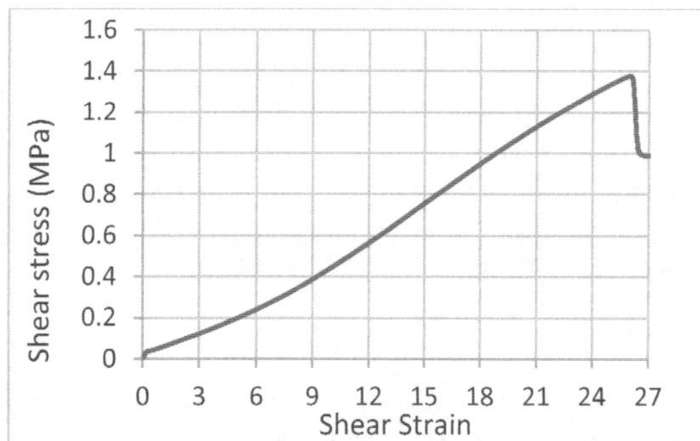


(b)

**Figure 2.** Cyclic tests at low speed (a) and monotonic shear test up to failure (b). Test with isotropic specimens and no magnetic field applied.

## References


*Schubert G and Harrison P (2015) Large-strain behaviour of Magneto-Rheological Elastomers tested under uniaxial compression and tension, and pure shear deformations. Polymer Testing 42. Elsevier Ltd: 122–134. DOI: 10.1016/j.polymertesting.2015.01.008.*

# Behaviours of Lithium-based Magnetorheological Grease under Triangular Quasi-static Test

**Huixing Wang[1], Yancheng Li[2,3*], Guang Zhang[1], Jiong Wang[1], and Jianchun Li[3]**

[1]*School of Mechanical Engineering, Nanjing University of Science and Technology, Nanjing, Jiangsu 210094, China*

[2]*College of Civil Engineering, Nanjing Tech University, Nanjing, Jiangsu 211816, China*

[3]*School of Civil and Environmental Engineering, University of Technology Sydney, Ultimo, NSW 2007, Australia*

[*]*yli@njtech.edu.cn and yancheng.li@uts.edu.au*

**Abstract**

This paper investigates the behaviour of lithium-based magnetorheological (MR) grease under the triangular quasi-static test. Three types of MR grease are prepared with weight fractions of carbon iron particles (CIP) as 30%, 50% and 70%, respectively. Quasi-static test of periodical triangular inputs, with various shear strain and strain rates, are employed to evaluate the performance of the MR greases, figure 1 and 2. Further evaluations are conducted by cross-checking the behaviour of the MR grease under various strain rate at a given max strain and the cases under various shear strains at a fixed strain rate.

## Introduction

MR grease is a new type of magneto-induced smart material composed of soft magnetic particles and grease matrix, which exhibits a transition state between traditional MR fluid (liquid state) and MR elastomer (solid state) in the absence of magnetic field (Ashour et al 1996; Shah et al 2013; Ginder et al 1996). In the case of vibration reduction of bridges and buildings, MR dampers are often work under periodic quasi-static loads. At this time, MR materials inside the damper are subjected to large deformation and quasi-static shearing. The field-induced properties of MR materials under quasi-static shear will determine the performance of MR damper used in bridges and buildings. Therefore, it is especially necessary to carry out the reaseach on quasi-static rheological properties of MR grease under large deformation to promote engineering application.

## Methods/Model Formulations

The quasi-static properties of MR grease were measured by a parallel-plate rheometer (type MCR 302, Anton Paar Co., Austria) equipped with magnetorheological module (Type MRD180, Anton Paar Co., Austria) and temperature-control accessory (Type F25, Julabo Technology Co, Germany) through the quasi-static monotonic shear and cyclic shear test. The sketch map of MCR 302 rheometer is shown in figure 1(a). In this experiment, MR grease is placed in the gap between the base mounted on worktable and parallel-plate connected to test head. The parallel-plate has a diameter of 20mm and its distance from the base is always set to 1mm.The magnetic field strength passing through the gap increase from 0 kA/m to 866 kA/m when regulating the coil current change from 0A to 5A.

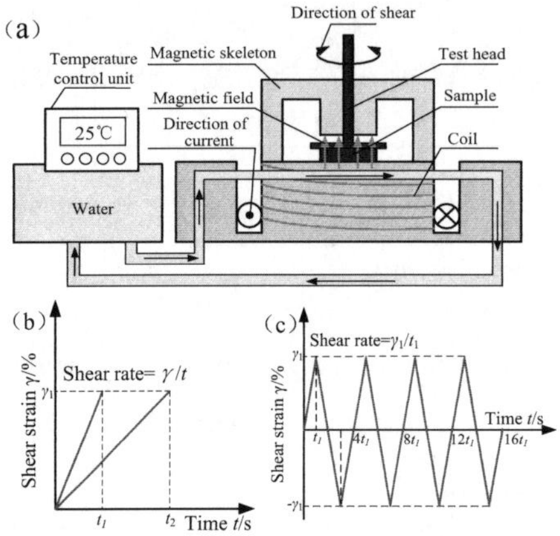


**Figure 1**. The sketch map of MCR 302 rheometer (a), the quasi-static monotonic shear test signal (b) and the quasi-static cyclic shear test signal (c)

## Results and Discussions

Some of the results are presented as the following and detailed discussion of the results will be presented in the full paper due to the limit of the extended abstract.

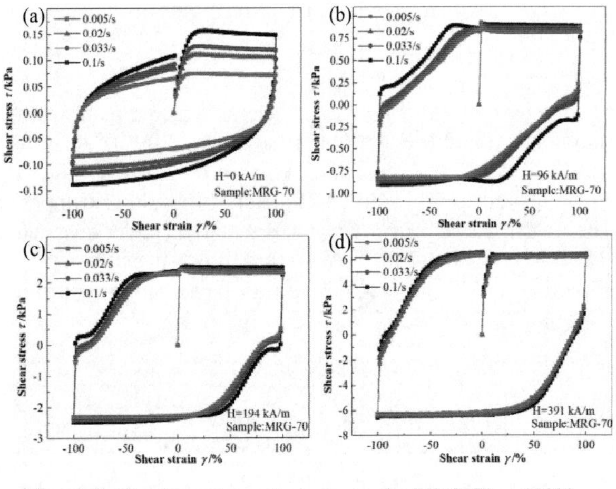


**Figure 2**. The shear stress-strain curves for MRG-70 at 100% strain under different loading conditions: (a)H=0 kA /m;(b)H=96 kA /m;(c)H=194 kA /m;(d)H=391 kA /m

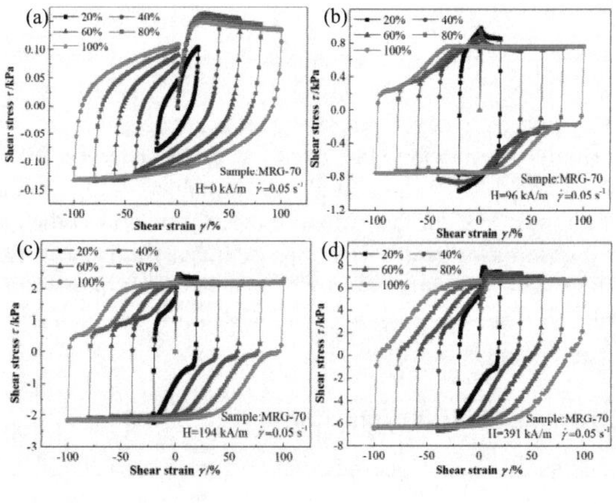


**Figure 3**. The stress-strain curves of MRG-70 at fixed strain rate with various shear strains

## References

*Ashour O, Rogers CA, Kordonsky W. (1996) Magnetorheological fluids: materials, characterization, and devices. Journal of Intelligent Material Systems and Structures 7(2): 123-30.*

*Shah K, Seong M-S, Upadhyay R, Choi S-B. (2013) A low sedimentation magnetorheological fluid based on plate-like iron particles, and verification using a damper test. Smart Materials and Structures 23(2): 027001.*

*Ginder J, Davis L, Elie L. (1996) Rheology of magnetorheological fluids: Models and measurements. International journal of modern physics b 10(23n24): 3293-303.*

# On the Role of Shape Factor on the Properties of Magnetorheological Elastomers in Compression Mode

**Hossein Vatandoost*, Masoud Hemmatian, Ramin Sedaghati, and Subhash Rakheja**

*Department of Mechanical, Industrial and Aerospace Engineering, Concordia University, Montreal, QC, H3G 1M8, Canada*

**Hossein.vatandoost@concordia.ca*

**Abstract**

In the present study, the dynamic properties of MREs are experimentally characterized in the compression mode considering variations in the shape factor. For this purpose, an experimental set up was designed to facilitate measurements of compression mode stress-strain behavior under controlled magnetic field. The experiment was designed to investigate the effect of shape factor of the MRE samples, defined as the ratio of the area subjected to an applied load to the force-free area, on the storage and loss moduli of the MRE under wide ranges of excitation frequency and external magnetic field intensity. The MRE samples were fabricated to realize shape factors ranging from 0.375 to 0.75, with fixed iron particle volume concentration of 30%. In the experiment, each MRE sample was positioned between two compression plates fixed to an optimally designed UI shaped electromagnet. Each specimen was subjected to strain amplitude from 2.5% to 20% at different frequencies, ranging from about 0.1 Hz to 30 Hz, while the magnetic field intensity was varied from 0 to 0.750 Tesla. The force and deflection responses of each sample were acquired and analyzed to determine the effect of shape factor on viscoelastic properties of the MRE as functions of the applied magnetic field and strain rate. The results suggested important effect of the shape factor on the distribution of stress in the vicinity of the fixed boundaries and thereby the viscoelastic properties. Experimental results were indicative of an increase in effective Young' modulus of MRE as the shape factor increases both in the absence and presence of external magnetic field. It was subsequently concluded that the sample's shape factor is a significant parameter which needs to be considered in characterizing properties of MREs.

**Introduction**

Magnetorheological elastomers (MRE) are categorized as the intelligent materials which exhibit variable/controllable mechanical properties in response to an external magnetic field. The rapid response to the applied magnetic field makes the MREs an ideal candidate for many engineering applications. Owing to MRE's attractive potentials, considerable efforts have been made toward their characterizations seeking for guidance on their fabrication and design for different applications. While shear properties of MREs and MRE based devices have been widely investigated (Dargahi et al, 2019, Johnson et al, 2017, Norouzi et al, 2017), very limited study has been conducted either theoretically or experimentally on characterization of MREs in compression mode (Vatandoost et al, 2017). Besides, the effect of shape factor on the MRE properties in the compression mode has been rarely investigated. The present study investigates the effect of shape factor on the behavior of the MREs operating in compression mode.

**Experiments**

Different MRE samples were fabricated to realize shape factors ranging from 0.375 to 0.75, with fixed iron particle volume concentration of 30%. Figure 1 shows the tests setup utilized to conduct the experiments in compression mode. The setup consists of lower and upper fixture which connected to the actuator and the load cell of a servo-hydraulic material testing machine (MTS), respectively. The Two identical MRE samples were placed between the "U" shape (lower fixture) and "I" shape (upper fixture) magnetic cores. A series of test were conducted on the MRE samples with different shape factors under different levels of applied flux density (0.0 and 600 mT) while the pre-strain and driving frequency were kept steady at 21% and 0.1 Hz, respectively. The obtained force-displacement hysteresis loops are presented in the following section and discussed.

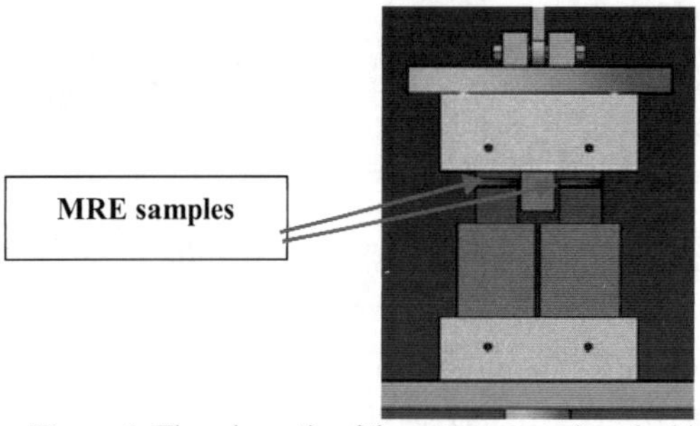


**Figure 1.** The schematic of the test setup equipped with the UI shape electromagnet.

### Results and Discussions

Figures 2(a) and 2(b) show the force-displacement hysteresis loops obtained at different levels of shape factor for magnetic flux density of 0 mT and 600 mT, respectively. By increasing the shape factor of MRE specimen, the hysteresis curves change from relatively linear elliptical to nonlinear asymmetric shape. Moreover, Figure 2 illustrates that both the slope and the area of the hysteresis loops, which respectively indicates the equivalent stiffness and damping of MREs, enhance by increasing shape factor. Therefore, results indicating strong dependence of the MRE properties on the shape factor both in the absence and presence of external magnetic field.

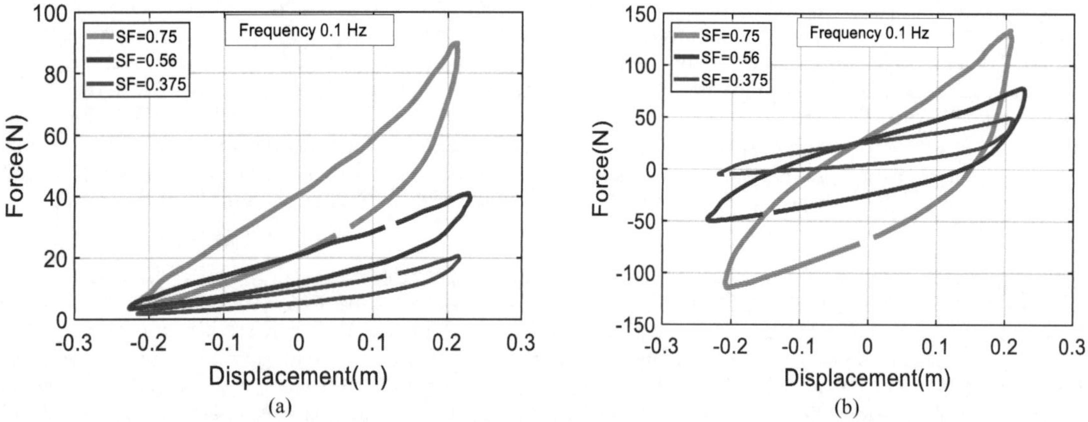


**Figure 2.** Hysteresis loops at Magnetic flux density of (a) 0mT, and (b) 600 mT

### References

*Dargahi, A., Sedaghati, R. & Rakheja, S. 2019. On the properties of magnetorheological elastomers in shear mode: Design, fabrication and characterization. Composites Part B: Engineering, 159, 269-283.*

*Johnson, N., Gordaninejad, F. & Wang, X. 2017. Dynamic behavior of thick magnetorheological elastomers. Journal of Intelligent Material Systems and Structures, 29, 183-193.*

*Norouzi, M., Gilani, M., Alehashem, S. M. S. & Vatandoost, H. 2017. Dynamic Characterization and Modeling of Isotropic Magnetorheological Elastomers Under Tensile-Compressive Loadings. IEEE Transactions on Magnetics, 53, 1-12.*

*Vatandoost, H., Norouzi, M., Alehashem, S. M. S. & Smoukov, S. K. 2017. A novel phenomenological model for dynamic behavior of magnetorheological elastomers in tension–compression mode. Smart Materials and Structures, 26, 065011.*

# Enhancement of MR-effect in Magnetorheological Elastomers through Bi-layer Composition: Theory and Validation

**Ali Alkhalaf, Amir Hooshiar*, and Javad Dargahi**

*Mechanical, Industrial, and Aerospace Engineering, Concordia University, Montreal, QC, Canada*

*s_hooshi@encs.concordia.ca

**Abstract**

In this study, an MRE−non-MRE bi-layer composition approach was postulated for enhancement of the MR-effect in magneto-rheological elastomers. To this end, a constitutionally necessary condition was derived for enhancing the MR-effect. Both single and bi-layer MREs were tested under compression with a transverse magnetic field. Experimental results validated the hypothesis and showed enhanced MR-effect in bi-layer MREs up to 42%.

## Introduction

Magnetorheological elastomers (MREs) are a class of smart materials with controllable mechanical properties. MREs are composed of ferrous particles diffused in a rubber medium. Interaction of the ferrous particles with the magnetic field and the elastomeric host matrix determines the mechanical properties of an MRE (Asadi Khanouki et al., 2019). Studies have shown that the volume fraction of iron particles and type of the elastomeric matrix affect the MR-effect, i.e. change in the elastic modulus of MREs (Dargahi et al., 2019). Studies have reported a positive correlation between iron particle content with the MR-effect in MREs. However, experimental results have shown a saturation volume fraction (40-45%) for iron particles. Therefore, the saturation is a barrier to achieving further MR-effect. Higher MR-effect is also desirable for better adaptability of MRE-based structures. In this study, a bi-layer MRE−non-MRE composite configuration was proposed and assumed to have a higher MR-effect than its constituent MRE. Afterward, a necessary constitutive condition was found for the elastic modulus of non-MRE component that necessarily would enhance the MR-effect in the composite.

## Material and Methods

To study the feasibility of enhancing MR-effect in an MRE through bi-layer composition, MRE (A), non-MRE material (B), and bi-layer composite (C) were assumed to exhibit a power-law stress-strain relationship (Eq. 1). This assumption could further be specified to an Ogden or Mooney-Rivlin solid. Also, it was assumed that initial shear modulus and absolute MR-effect of MRE A are a priori knowledge.

$$\sigma_{A,B,C} = \mu_{A,B,C}\lambda_{A,B,C}^{\alpha} \tag{1}$$

Where, $\mu$ , $\alpha$, $\sigma$, and $\lambda$ were the fitting parameters, engineering stress and, stretch ratio. Previous studies have postulated and validated $\alpha = 1.3$ for EcoFlex silicone-based rubbers in compression. Also, the required condition for the enhanced MR-effect was defined as:

$$\mu'_C - \mu_C > \mu'_A - \mu_A, \tag{2}$$

where, $\mu_A$ and $\mu_C$ were the initial shear moduli of MRE A and bi-layer composite at zero magnetic field, while $\mu'_A$ and $\mu'_C$were the elastic moduli of elastomers under magnetic field. Considering the structure of composite as depicted in Fig. 1, the effective initial and final shear modulus of composite C was obtained:

$$\mu_C = 2^{\alpha}\left(\mu_A^{\frac{-1}{\alpha}} + \mu_B^{\frac{-1}{\alpha}}\right)^{-\alpha} \tag{3}$$

$$\mu'_C = 2^{\alpha}\left(\mu_A'^{\frac{-1}{\alpha}} + \mu_B^{\frac{-1}{\alpha}}\right)^{-\alpha} \tag{4}$$

Therefore, substituting Eq. 3 and Eq. 4 in Eq. 2, and the explicit form of the necessary condition was derived:

$$\left(\mu_A'^{\frac{-1}{\alpha}} + \mu_B^{\frac{-1}{\alpha}}\right)^{\alpha} - \left(\mu_A'^{\frac{-1}{\alpha}} + \mu_B^{\frac{-1}{\alpha}}\right)^{\alpha} - (\mu'_A - \mu_A)\left(\frac{\left(\mu_A'^{\frac{-1}{\alpha}} + \mu_B^{\frac{-1}{\alpha}}\right)\left(\mu_A^{\frac{-1}{\alpha}} + \mu_B^{\frac{-1}{\alpha}}\right)}{2}\right)^{\alpha} > 0 \tag{5}$$

Numerical analysis showed that there exists a unique minimum $\mu_B$ which ensures the satisfaction of the condition. To solve the equation for the required $\mu_B$, shear modulus of a similar MRE with 40% iron particle content reported in (Dargahi et al., 2019) was used, i.e. $\mu_A = 32.66$kPa and $\mu'_A = 214.21$kPa. Newton-Raphson numerical solution revealed a required $\mu_B^{min} = 285.28$kPa. To validate the hypothesis, a particle saturated MRE with 45% volume-fraction was fabricated. MRE was composed of 45% volume fraction of EcoFlex 20-00 (SmoothOn Inc., PA, US) silicon rubber, 45% carbonyl iron particles (CIPs), 5% slacker and 5% thinner. Also, Dragon Skin 30 silicon rubber (SmoothOn Inc., Pa, US) was selected as the non-MRE component. Studies have reported a shear modulus of 370kPa for Dragon Skin 30. Therefore, this material would pass the criterion in Eq.5. Also, two sets of parallel permanent magnets were used to apply a transverse magnetic field (0-365mT). Three cylindrical samples of each material (n=9, 29×12.55mm) were tested in quasi-static sinusoidal compression with a strain amplitude of 25% and the frequency of 0.1Hz. Elastic moduli of samples were obtained following ISO 7743:2017.

## Results

From the stress strain-curve, the compressive elastic modulus of elastomer B was found to be 3635kPa. Substituting $\mu_B = 1211$kPa in Eq. 5 Also, elastic moduli of MRE A and composite C were 281kPa and 388kPa at zero magnetic field, and 873kPa and 1370kPa at 365mT. Therefore, MRE C exhibited an MR-effect of 253%, while its MR component, MRE A only showed an MR-effect of 211%. Fig.1(b) depicts vaa riation of compressive elastic moduli of MRE A, elastomer B and composite C. MR-effects were approximately half as reported in (Dargahi et al., 2019), the reason might be due to the transverse application of the magnetic field.

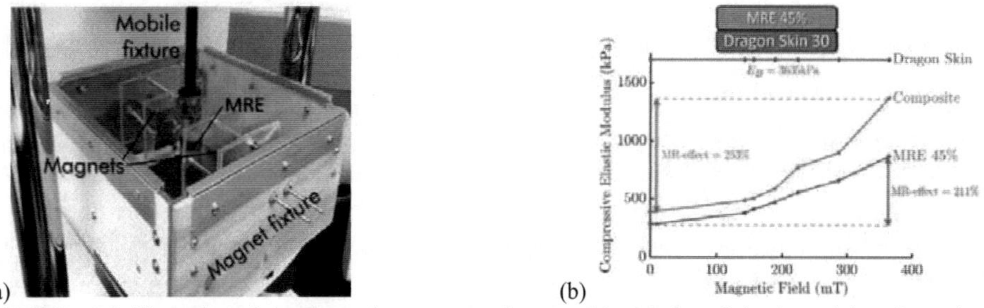


(a) (b)

**Figure 1.** (a) Schematic illustration of samples under a compression test. (b) variation of elastic modulus of samples versus magnetic field.

## Conclusions

In this study, a bi-layer MRE −non-MRE composite structure was postulated and validated. Surpassing the minimum required elastic modulus for the non-MRE component was theoretically predicted and experimental results validated the hypothesis. Although the MRE component of the composite in this study was saturated with CIPs, the MR-effect was successfully enhanced through the bi-layer composition.

## References

*Asadi Khanouki M., Sedaghati R. and Hemmatian M. (2019) Characterization of Magneto-Mechanical Properties and Quasi-Static Physical Modelling of MR Elastomers. In AIAA Scitech 2019 Forum (p. 1114).*

*Dargahi A., Sedaghati R. and Rakheja S. (2019) On the properties of magnetorheological elastomers in shear mode: Design, fabrication and characterization. Composites Part B: Engineering, 159, pp.269-283.*

*ISO 7743:2017 (2017) Rubber, vulcanized or thermoplastic -- Determination of compression stress-strain properties.*

# Dynamic Characterization of Magnetorheological Elastomers based on Hard Magnetic Particles under Harmonic Loading

**Nader Mohseni***, **Masoud Hemmatian, and Ramin Sedaghati**

*Department of Mechanical, Industrial and Aerospace Engineering, Concordia University, Montreal, QC, H3G 1M8, Canada*

**n_mohse@encs.concordia.ca*

**Abstract**

In this study, a magnetorheological elastomer based on hard magnetic particles (H-MREs) was fabricated and the viscoelastic properties of the H-MREs were further investigated under various levels of applied magnetic flux densities. Both positive and negative magnetic fields ranging from -0.2 T to 0.4 T, were applied and the dependence of storage and loss moduli on the strain amplitude were evaluated while the frequency was maintained steady at 1 Hz. Bi-directional variation of the storage moduli with change in applied magnetic flux density offers the potential of the HMRE in vibration control and other applications need negative stiffness or wide range of change in stiffness.

## Introduction

Magnetorheological elastomers (MREs) are an emerging class of smart materials that can change their properties in a reversible manner under the application of an external magnetic field (Dargahi et al, 2017). MREs can be used as variable stiffness and variable damping components in different mechanical devices. MREs consist of micron-sized iron particles arranged or suspended within an elastomer matrix. Soft magnetic particles, such as carbonyl iron (CI) powders, were widely used in MR materials, because of their high saturation magnetization and low remanence. Recently, particles of high-coercivity ferromagnetic materials, or hard-magnetic materials, such as neodymium-iron-boron (NdFeB) have been embedded in soft materials (Zhao et al, 2017). The high remnant characteristics of hard-magnetic materials allow them to retain high residual magnetic flux density even in the absence of magnetic fields once they are magnetically saturated (Wen et al, 2017). Most of the previous studies on H-MREs were mainly focused on the performance under a positive applied magnetic field. In the present study, both positive and negative magnetic fields are considered and their impact on the viscoelastic properties of the HMRE were investigated.

## Experiments

### Sample Preparation

The raw HMREs include two basic components: unmagnetized NdFeB particles (50 μm) andthe silicone rubber (Ecoflex 00-50) acts as the matrix. First, the NdFeB particles and the silicone rubber were mixed thoroughly for 5 min. into a homogenous mixture. Then, the mixture was put into vacuum chamber under -28 inHg for 5 min. for degassing. Afterward, the mixture was poured into a transparent plexi-glass mold and left for 24 hours at room temperature to cure. Finally, the cured material was cut into 20 mm diameter sample and magnetized using a double-yoke adjustable electromagnet under 2.7 T.

### Dynamic Characterization

The dynamic viscoelastic properties of the H-MRE with 15% volume fraction of hard particles were measured using a rotary rheometer (Discovery HR-3, TA instrument) equipped with magnetorheology accessory. The test sample (20 mm diameter and 2 mm thickness) was placed between the rotating disk and the base. The applied magnetic field provided by the magnetorheology accessory and applied perpendicular to the sample. The applied magnetic field could be controlled in two orientations. The orientation of the magnetic field is defined as positive when it was in the same direction as the direction of the inner particles' initial residual magnetization. The tests were performed using the dynamic oscillatory mode by sweeping the strain amplitude in the range of 0.01% to 10% while the frequency and temperature maintained steady at 1 Hz and 25 °C, respectively. The tests have been done under various levels of positive and negative applied magnetic fields and the dependency of the storage and loss moduli of the HMRE on the strain amplitude and applied magnetic flux density were investigated. It should be mentioned that 30 N axial load applied on the sample before initiating the tests.

**Results and Discussions**

As can be seen from Figure 1, the storage modulus increase with an increase of the applied magnetic flux density. Moreover, the shear storage modulus is independent from the strain amplitude for strain amplitude less than 0.1 % and further increase of the strain amplitude decreases the storage modulus. This phenomenon likely due to the change in the perfectly aligned iron particles owing to strain amplitude. It is also clear than unlike the soft MREs, applying magnetic flux density in the direction opposite to the orientation of the magnetization reduces the storage modulus of the HMRE. Thus, it is possible to make the HMRE softer or stiffer by applying the negative or positive magnetic field, respectively. This behavior offers the opportunity to use HMREs in different vibration control applications in which both negative and positive variation of the stiffness is required.

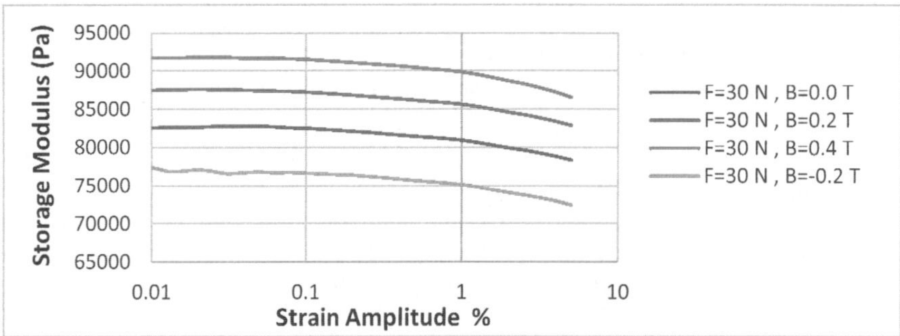


**Figure 1.** Dependence of storage modulus on the strain amplitude at the frequency of 1 Hz for various applied magnetic flux densities.

**References**

*Dargahi, A., Sedaghati, R. & Rakheja, S. 2017. Fabrication, Characterization and modeling of magnetorheological elastomers.*

*Wen Q, Yu Wang Y & Gong X, The magnetic field dependent dynamic properties of magnetorheological elastomers based on hard magnetic particles, Smart Mater. Struct. 26 (2017)*

*Zhao W, Pang H, Gong X, Novel Magnetorheological Plastomer Filled with NdFeB Particles: Preparation, Characterization, and Magnetic−Mechanic Coupling Properties (2017).*

# A Linear Model for Magnetoactive Elastomers

**Alireza Beheshti[*], Ramin Sedaghati, and Subhash Rakheja**

*Department of Mechanical, Industrial and Aerospace Engineering, Concordia University, Montreal, QC, Canada H3G 1M8*

[*] *a_behesh@encs.concordia.ca*

**Abstract**

The present work is centered on small deformation analysis of isotropic magnetoactive elastomers (MAEs). By taking into account the governing mechanical and magnetic equations for a flexible MA medium, the corresponding equations are expanded considering three separate states involving material, initial and spatial configurations. First a general material model for finite deformation of MAEs is proposed and then a linearization scheme is carried out to develop a constitutive equation for small deformation of MAEs. A rheometer equipped with an electromagnet device has been utilized to characterize fabricated isotropic samples with 15% volume fraction under various magnetic field intensities. The experimental results have then been utilized to estimate the material constant parameters.

## Introduction

MAE is a class of smart materials for which mechanical behaviors alter by the application of an external magnetic field. By employing the theory of magnetoelastic solids of Brown (1966) and making use of a linearization scheme, Pao and Yeh (1973) presented a theory for static small-deformation analysis of ferromagnetic solids by splitting the magnetic parameters in the deformed configuration into two constituents including rigid-body and perturbation status. Later, by adopting the same scheme for separation of magnetic variables, Hutter and Pao (1974) proposed a theory for dynamic small-strain analysis of a medium for which magnetic, electrical and thermal properties are considered. Pao (1978) presented two valid formulations for moving deformable solids and then, by taking into account the first and second law of thermodynamics, a constitutive equation in polynomial form was established for both theories.

## Methods/Model Formulations

In the first step, the local form of the basic laws in the continuum mechanics involving the balance law of linear and angular momentum for a magnetoactive medium which is under mechanical and magnetic loadings are introduced respectively as follows (Pao, 1978):

$$\nabla_{\mathbf{x}}.\boldsymbol{\sigma} + \mathbf{f} + \bar{\mathbf{f}} = \rho\frac{d\mathbf{v}}{dt}, \quad \frac{1}{2}(\boldsymbol{\sigma} - \boldsymbol{\sigma}^T) + \boldsymbol{\eta} = 0 \tag{1}$$

where $\mathbf{v}$ stand for velocity of a particle and $\boldsymbol{\sigma}$ , $\bar{\mathbf{f}}$ and $\mathbf{f}$ is the so-called Cauchy stress tensor, the mechanical and magnetic body force per unit volume. Additionally, $l_i = \varepsilon_{ijk}\eta_{jk}$ where $l_i$ is body couple per unit volume due to electromagnetic field. Furthermore, building upon the theory developed by Pao (1978) the magnetic body force and the body couple for the static case have the following forms:

$$\mathbf{f} = (\nabla_{\mathbf{x}}\mathbf{b}).\mathbf{m}, \quad \mathbf{l} = \mathbf{m} \times \mathbf{b} \tag{2}$$

In order to find the magnetic parameters inside the material, the Maxwell's equations should be solved. Thus, for the static case, we can write $\nabla_{\mathbf{x}}.\mathbf{b} = 0, \nabla_{\mathbf{x}} \times \mathbf{h} = \mathbf{0}$, where $\mathbf{b}$, $\mathbf{h}$ and $\mathbf{m}$ are the magnetic field induction, magnetic field strength and magnetization. The capital letters are used for the magnetic variables in the initial configuration, whereas the magnetic parameters are written in small letters for the spatial configuration. From the initial to deformed configuration, there is a variation in the magnetic parameters due to the mechanical deformation of the body. Thus, we decompose the spatial magnetic parameters into two constituents involving the initial fields $\{\mathbf{B}, \mathbf{H}, \mathbf{M}\}$ and the corresponding variations $\{\hat{\mathbf{b}}, \hat{\mathbf{h}}, \hat{\mathbf{m}}\}$ as:

$$\mathbf{b} = \mathbf{B} + \hat{\mathbf{b}}, \qquad \mathbf{h} = \mathbf{H} + \hat{\mathbf{h}}, \qquad \mathbf{m} = \mathbf{M} + \hat{\mathbf{m}} \tag{3}$$

By employing a linearization scheme on the stress tensor for incompressible MAEs can be found to be:

$$\bar{\boldsymbol{\sigma}} = p\mathbf{I} + 2\mu^{*}\mathbf{e} + 2\gamma B_k \hat{b}_k \mathbf{I} \tag{4}$$

## Results and Discussions

For small-deformation of a right circular cylinder, the non-vanishing strains are $e_{13} = -\kappa x_2/2$: and $e_{23} = \kappa x_1/2$ where $\kappa$ is the angle of twist per unit length. With the knowledge of the distribution of strain in the cylinder, and then by making use of the constitutive equation, the shear components of the tensor can be estimated as $\sigma_{13} = 2\mu^{*}e_{13}$ and $\sigma_{23} = 2\mu^{*}e_{23}$. The stress field should satisfy Eq. (1). In addition, based on the experiment carried out on the fabricated cylindrical MAEs, the nominal shear modulus has been reported in Figure for different values of magnetic fields for 15% MAE. It should be pointed out that the nominal shear modulus obtained from the torsion test is indeed the ratio $\mu^{*}=T/\kappa I$ where $T$ is the applied torque and $I$ is the polar moment of inertia of the cross section.

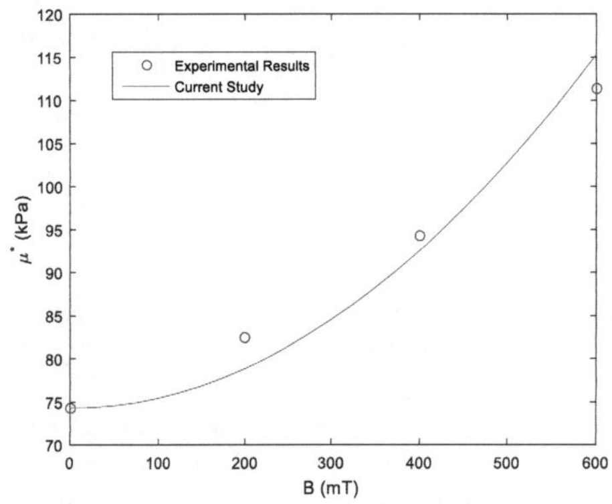


**Figure 1**. Nominal shear modulus versus magnetic field.

Figure 1 implies that the nominal shear modulus is a function of the magnetic field. Accordingly, the nominal shear modulus can be represented by $\mu^{*} = \mu(1 + \eta I_4)$, where μ and η are material constants. Considering Eq.(6), it is concluded that the value of μ is 74.394 kPa and η is 1.532 $1/T^2$. It is noted that the constant μ is the so-called shear modulus in solid mechanics.

## References

*BROWN, W. F. 1966. Magnetoelastic Interactions. Springer-Verlag Berlin. Heidelberg New York.*

*HUTTER, K. & PAO, Y. H. 1974. A dynamic theory for magnetizable elastic solids with thermal and electrical conduction. Journal of Elasticity, 4, 89-114.*

*PAO, Y.-H. 1978. Electromagnetic forces in deformable continua. Cornell University, Ithaca, New York 209-305.*

*PAO, Y.-H. & YEH, C.-H. 1973. A LINEAR THEORY FOR SOFT FERROMAGNETIC. International Journal of Engineering Science, 11, 415-436.*

# On the Effects of Pre-strain on Dynamic Behavior of Magnetorheological Elastomers in Compression Mode

**Hossein Vatandoost***, **Masoud Hemmatian, Ramin Sedaghati, and Subhash Rakheja**

*Department of Mechanical, Industrial and Aerospace Engineering, Concordia University, Montreal, QC, H3G 1M8, Canada*

**Hossein.vatandoost@concordia.ca*

**Abstract**

In practical engineering applications, MREs are typically pre-deformed and operate with geometrical constraints, thereby inducing different levels of pre-strain. In the present study, the dynamic properties of MREs are experimentally characterized in the compression mode considering variations in the pre-strain. For this purpose, an experimental set up was designed to perform measurements in compression mode under controlled magnetic field. The experiments were designed to investigate the effect of pre-strain on the equivalent stiffness and damping of the MRE under wide ranges of excitation frequency and external magnetic field intensity. Each MRE specimen was subjected to various levels of pre-strain and strain amplitudes ranging from 6% to 21% and 2.5% to 20%, respectively, at different excitation frequencies (0.1-30 Hz), under different intensities of the applied magnetic field varied from 0 to 0.750 T. The force and displacement responses were acquired and analyzed to determine the effect of pre-strain on viscoelastic properties of the MRE as a function of the applied magnetic field. The results suggest a significant effect of the pre-strain on the viscoelastic properties of MRE both in the absence and presence of external magnetic field.

**Introduction**

Magnetorheological elastomers (MRE) are a class of functional intelligent materials, which consist of solid polymeric matrix and magnetic particles, and exhibit variable/controllable mechanical properties in response to an external magnetic field. Owing to their fast and reversable response to the applied magnetic field, the MREs are considered ideal candidates for many engineering applications (Li et al, 2014). Considering MRE's attractive potentials, substantial efforts have been implemented toward their characterizations looking for designing MRE-based devices for applications in noise and vibration suppression. While shear properties of MREs and MRE based devices have been extensively scrutinized (Dargahi et al, 2019, Norouzi et al, 2016, Norouzi et al, 2017, Gong et al, 2005), a fewer studies have been conducted on characterization of MREs in compression modes (Vatandoost et al, 2017). Furthermore, the effect of pre-strain on the MRE properties in the compression mode has been rarely investigated. The present study aims to investigate the effect of pre-strain on the viscoelastic properties of MRE, operating in compression mode, under various levels of applied flux density.

**Experiments**

Isotropic MREs were manufactured in the laboratory with 30% particle volume fraction. The compression test setup consists of a servo-hydraulic material testing machine (MTS) and a tailor-made UI-shaped electromagnet device including three winding coils and UI-shaped magnetic core as shown in Figure. The lower fixture was attached to the actuator and upper fixture was fixed to the load cell. Two identical MRE specimens were positioned between the U-shaped and I-shaped magnetic cores. Since the MRE samples are tested under a pre-compressed state, the need for any adhesive is eliminated. A series of compression training set data tests were performed on different MRE samples by varying the magnetic flux density, strain amplitude, and pre-strain. Magnetic flux densities were varied from zero to 750 mT whereas frequency of 1 Hz was maintained. Different force-displacement hysteresis curves obtained under different levels of pre-strain are discussed in the next section.

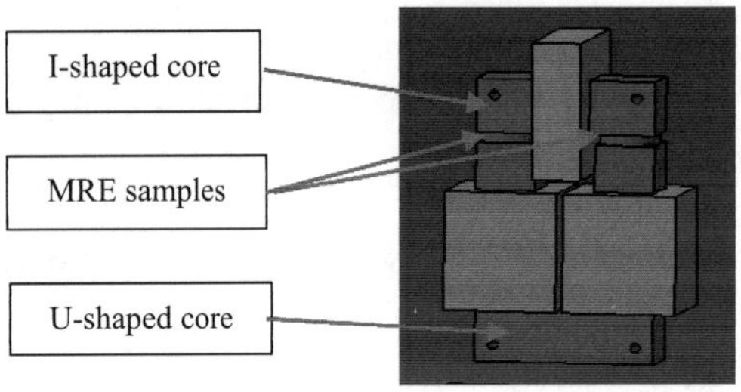


**Figure 1.** The schematic of the test setup equipped with the UI-shaped electromagnet device.

## Results and Discussions

Figures 2(a) and 2(b) present the Force-displacement hysteresis loops are obtained at different level of pre-strain in absence and presence of 300 mT magnetic flux density. It is clear in this figure that by increasing the pre-strain, the hysteresis curves move upward represent pre-strain stiffening phenomenon. Moreover, Figure 2 illustrates that both the slope and the area of the hysteresis loops, which respectively indicate the equivalent stiffness and damping of MREs, enhance by increasing the pre-strain. Results revealed 191.5% and 35.9% increase in equivalent stiffness when pre-strain increases from 6% to 21% in absence and presence of 300 mT magnetic flux density, respectively, thereby suggesting that the pre-strain stiffening phenomenon is more pronounced in lower magnetic field.

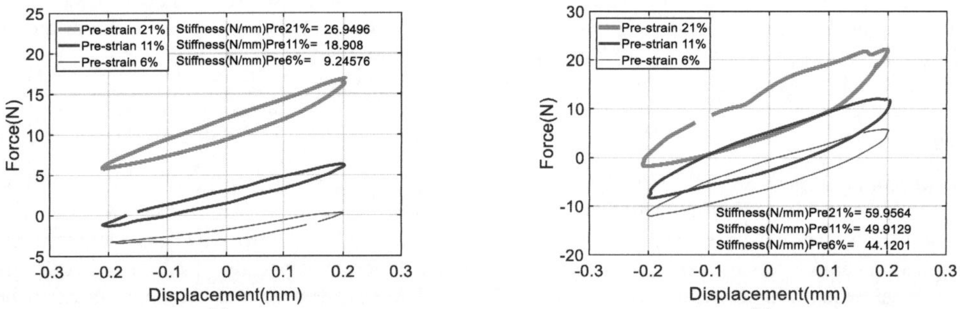


(a) (b)

**Figure 2**. Hysteresis loops of the MRE at different levels of pre-strain in presence of (a) zero and (b) 300 mT applied magnetic flux density (f = 1 Hz and $\gamma$ = 2.5 %).

## References

*Dargahi, A., Sedaghati, R. & Rakheja, S. 2019. On the properties of magnetorheological elastomers in shear mode: Design, fabrication and characterization. Composites Part B: Engineering, 159, 269-283.*

*Gong, X. L., Zhang, X. Z. & Zhang, P. Q. 2005. Fabrication and characterization of isotropic magnetorheological elastomers. Polymer Testing, 24, 669-676.*

*Li, Y., Li, J., Li, W. & Du, H. 2014. A state-of-the-art review on magnetorheological elastomer devices. Smart Materials and Structures, 23, 123001.*

*Norouzi, M., Gilani, M., Alehashem, S. M. S. & Vatandoost, H. 2017. Dynamic Characterization and Modeling of Isotropic Magnetorheological Elastomers Under Tensile-Compressive Loadings. IEEE Transactions on Magnetics, 53, 1-12.*

*Norouzi, M., Sajjadi Alehashem, S. M., Vatandoost, H., Ni, Y. Q. & Shahmardan, M. M. 2016. A new approach for modeling of magnetorheological elastomers. Journal of Intelligent Material Systems and Structures, 27, 1121-1135.*

*Vatandoost, H., Norouzi, M., Alehashem, S. M. S. & Smoukov, S. K. 2017. A novel phenomenological model for dynamic behavior of magnetorheological elastomers in tension–compression mode. Smart Materials and Structures, 26, 065011.*

# The Influence of GFRP Web Reinforcement on the Structural Behaviour of Deep Beams

**Fawzi Latosh* and Ashutosh Bagchi**

*Department of Building, Civil & Environmental Engineering, Concordia University, Sir George Williams Campus 1455 De Maisonneuve Blvd. W. Montreal, Quebec, Canada H3G1M8*

**fawzi.latosh@concordia.ca*

**Abstract**

Fiber Reinforced Polymer (FRP) materials are a strong material, light and free of corrosion problem and therefore being used alternatively of steel reinforcements in concrete structural elements. Many studies have investigated the effectiveness of longitudinal FRP bars, but not many studies are available on the efficiency of the FRP web reinforcements in concrete deep beams. A set of four Glass Fiber Reinforced Polymer (GFRP) Reinforced Concrete deep beam with a shear span-to-depth ratio equal to one and a different aspect ratio of web reinforcement were tested to observe their behavior. The study demonstrates the importance of the web reinforcements in controlling the overall behavior of reinforced concrete deep beams for example the mid-span deflection, crack width, failure modes and ultimate strengths. In addition, this study developed a new equation to predict the contribution of the FRP web reinforcements to the ultimate shear capacity of FRP-reinforced concrete deep beams.

**Introduction**

The Strut and Tie model (STM) has been adopted by the modern codes to design the deep beams in a more effective way. The STM provides a rational and acceptable theory for the design of deep beams, which generally agrees well with the results of experimental studies. It should be noted that the implementation of the STM models as provided in different codes (ACI 318-14 2014, Eurocode 2004 and CAN/CSA-S806-12 2012) are differ from each other. Some differences exist among the codes in the implementation of STM, particularly in determining the amount of web reinforcement and the shape of the struts. Earlier works by Smith and Vantsiotis 1983, Tan et al 1997, Shin et al1999, Kong et al 1970 and Rogowsky et al1986 on the effect of web reinforcement on the various aspects of deep-beam (conventional) behaviour showed that web reinforcement have a significant effect on the mid-span deflection, crack-width, failure modes and ultimate strengths. Many researchers have observed that web reinforcement increases beam stiffness, and that this influence becomes significant according to the arrangement and amount of web reinforcement, and on the L/d and a/d ratios. Despite the divergent views in determining the effect of web reinforcements on crack control, the majority has concluded that the effect of web reinforcements on crack width and crack control is akin to its effect on beam stiffness. Most of the researchers clarified that beams with web reinforcement exhibit the same modes of failure as compared to the beams without web reinforcement. For beams with web reinforcement the ultimate shear strength of deep beams is slightly increased as compared to those without; this contribution of shear reinforcement has not yet been determined as it has in conventional deep beams.

**Experimental Plan**

Full detailing of specimen dimensions and reinforcement are illustrated in Figures (1).

**Results and Discussions**

The formation of the first flexural crack at mid-span for four beams appears at approximately 17 to 24% of the ultimate load according to the amount of the web reinforcement. The failure crack patterns within the four beams are similar; however, the beams without web reinforcement are more extensively cracked than the beams with web reinforcement. The mid-span strain response of the main GFRP longitudinal reinforcements at different loading stages for specimens are shown in Figure (2).

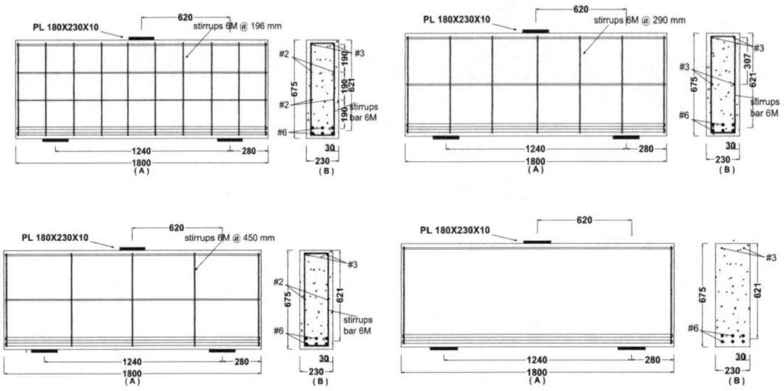


**Figure 1**. T (A) elevation (B) cross section of the four specimens

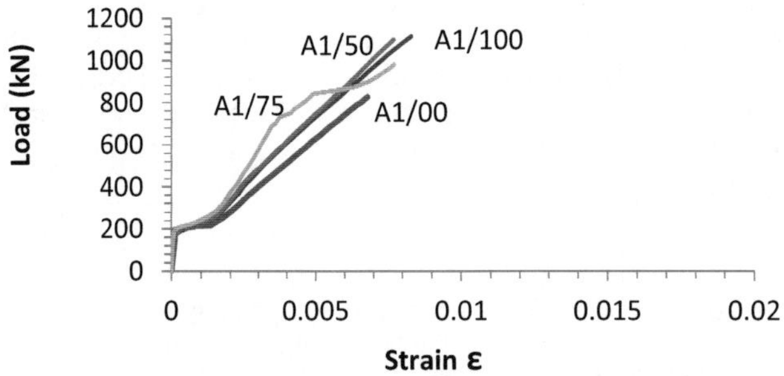


**Figure 2**. Load –Strain response in the middle of the main tension GFRP reinforcement of specimens

## References


*European Standard EN1992-1-1 (2004) Euro code 2: Design of Concrete Structures – Part 1-1: General Rules and Rules for Buildings. European Committee for Standardization, CEN*

*ACI Committee 318, (2014) Building Code Requirements for Structural Concrete (ACI 318-08) and Commentary (318R-08). American Concrete Institute, Farmington Hills, MI, 430*

*CAN/CSA S806-12 (2012) Design and construction of building components with fiber-reinforced polymers. Canadian Standards Association, Mississauga, Ontario, Canada, 208*

*Smith, K. N., and Vantsiotis, A. S. (1983) Shear strength of deep beams. ACI Structural Journal, V. 79, No. 3, pp. 201–213.*

*Tan, K. H., Kong, F. K., Teng, S., and Weng, L. W.(1997) Effect of Web Reinforcement on High Strength Concrete Deep Beams. ACI Structural Journal, V. 94, No. 5, pp. 572-581.*

*Shin, S. W., Lee, K. S., Moon, J., and Ghosh, S. K. (1999) Shear strength of reinforced high-strength concrete beams with shear span-to-depth ratios between 1.5 and 2.5. ACI Structural Journal, V. 96, No. 4, pp. 549-556.*

*Kong, F. K., Robins, P. J., Cole, D. F. (1970) Web Reinforcement Effects on Deep Beams. ACI Structural Journal, V.67, No. 12, pp. 1010-1018.*

*Rogowsky, D. M., MacGregor, J. G., and Ong. S. Y.(1986) Tests of Reinforced Concrete Deep Beams. ACI Structural Journal, V. 83, No. 4, pp. 614-623.*

# Design and Analysis of the Mechanical Properties of 3D Printed Sinusoidal Structures with Auxetic Characteristics

**Diego A. Quevedo-Moreno, Armando Román-Flores*, Enrique Cuan-Urquizo, Marcos D. Moya-Bencomo, Adriana Vargas-Martínez**, and Ricardo Ramírez-Mendoza**

*Tecnologico de Monterrey, School of Engineering and Sciences, Mexico*

*armando.roman@tec.mx, **adriana.vargas.mtz@tec.mx

**Abstract**

The auxetic materials are materials that have abnormal behavior; they have a Poisson's coefficient with a negative sign, unlike conventional materials, which is positive. The Poisson`s coefficient determines the relationship between the transverse deformation and the longitudinal deformation of a material when a tension or compression force is applied (Crespo and Montáns, 2018). With auxetic materials, when the longitudinal deformation is positive, the transverse deformation is positive, so we have an expansion of the material (Zhang et al., 2018). This unusual feature makes the study and development of auxetic materials striking because it can have potential applications in various areas such as medical, automotive, armament, sports, among others (Günaydin et al., 2017). Because of the above, in this work, the development of 3D printed auxetic structures with sinusoidal patterns is presented. Six different types of specimens were designed, where the amplitude of the sinusoids was varied to give shape to the auxetic structure. Tension tests were performed in each of them to characterize them.

**Methods/Model Formulations**

The specimens designed with the auxetic structures were modeled mathematically in MATLAB, with vertical and horizontal sinusoidal functions. The test pieces have 5 extrusion layers of 1.75mm PLA filament material. Six different types of specimens were designed, where the amplitude of the sinusoids was varied to give shape to the auxetic structure (figure 1).

**Figure 1.** Amplitude variation of the test tubes from 0 to 2.5mm with a step of .5mm

To characterize them, tension tests were performed with a universal machine (MTS Electromechanical Universal Testing Systems), photos are taken every 2.5lbf applied on the test piece to later measure the axial and lateral deformation with Digital Image Correlation (DIC GOM Correlate 2017).

To perform the analysis of the Poisson's coefficient and Young's modulus, the elastic zone of each structure was calculated during the stress test. This area was estimated by making a straight line with slope equal to the linear regression of the data obtained with compensation of .2%.

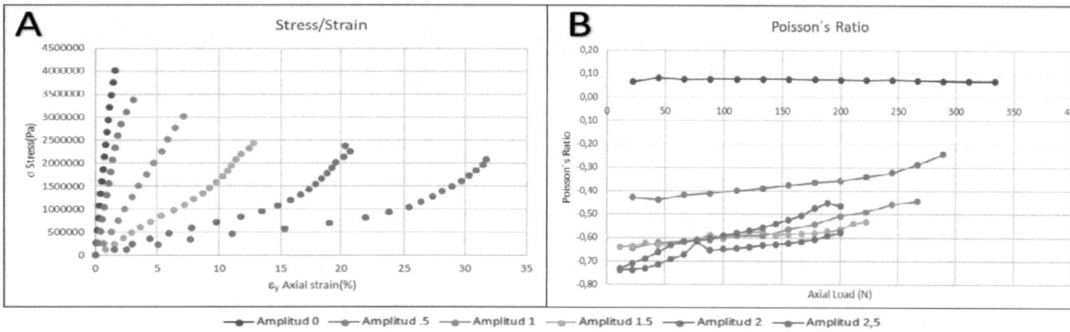


**Figure 2.** A) Stress/Strain Graph of the tested specimens, B) Poisson`s Ratio/Axial Load Graph of the tested specimens.

**Table 1.** Mechanical Properties

| Amplitude (mm) | Poisson´s Ratio | Young Modulus (Pa) |
|---|---|---|
| 0 | 0,07 | 292606740,01 |
| 0,5 | -0,40 | 1359242,06 |
| 1 | -0,56 | 414478,25 |
| 1,5 | -0,62 | 158789,44 |
| 2 | -0,70 | 75631,40 |
| 2,5 | -0,67 | 42105,82 |

## Results and Discussions

The proposed sinusoidal structures have a negative Poisson coefficient, except for the one of amplitude 0. The higher the amplitude, the easier it is to deform the structure, and the Poisson coefficient is more negative, but there is a lower Young's modulus. During stress tests, when going over the structural elastic system, the slope increases in the stress curve against deformation; this occurs because the structure tries to return to the typical properties of the material. The same happens with the Poisson coefficient, when subjecting the structure to significant stresses and when decreasing the amplitude of this by the generated deformation the Poisson coefficient tends to become positive.

## References

*Crespo, J, and Montáns, F. (2018). A continuum approach for the large strain finite element analysis of auxetic materials. In International Journal of Mechanical Sciences, 135, pp. 441-457.*

*Günaydin, K., Eren, Z., and Scarpa, F. (2017). Experimental investigation of auxetic structures subjected to quasi-static axial load. In International Conference on Recent Advances in Space Technologies (RAST), pp. 7-10.*

*Zhang, J., Lu, G., Ruan, D., and Wang, Z. (2018). Tensile behavior of an auxetic structure: Analytical modeling and finite element analysis. In International Journal of Mechanical Sciences, 136, pp. 143-154.*

**Gap in pagination due to unavailable papers.**

**Pages 147-148**

# Tuning the Hygro-mechanical Response of Paper-based Systems using Glycerol

**Isaias Cueva-Perez, Roque Alfredo Osornio-Rios, Aurelio Dominguez-Gonzalez, and Angel Perez-Cruz***

*Mecatronica/Facultad de Ingenieria, Campus San Juan del Rio, Universidad Autonoma de Queretaro Rio Moctezuma 249, Colonia San Cayetano, 76807 San Juan del Rio, Queretaro, Mexico*

**angel.perez@uaq.mx*

## Abstract

Paper have been recently used as substrate material in the development of devices such as sensors and actuators. The physical properties of paper have improved the ease of fabrication of this kind of devices, thus opening novel possibilities such as the tuning of the mechanical properties with different purposes. In this work, a tunable method using glycerol is presented. Different sample configurations are fabricated and experimentally tested in order to study the hygro-mechanical response of paper due to its interaction with glycerol. The results suggest that the combination of the paper sample configurations and the glycerol can lead to the development of paper-based systems that can be tuned specifically to a desired frequency.

## Introduction

Paper-based mechanical structures have offered novel possibilities regarding the development of sensors and actuators due to its multiple advantages such as low cost, biodegradability, disposability, flexibility, capillarity, low weight and ease of fabrication (Wang et al., 2018). Due to these properties, it is possible to implement tuning methods in order to modify the dynamic response of paper with relative ease. In this work, a tunable method using a glycerol-based solution is presented. Different configurations of the paper-based systems are fabricated and experimentally tested in order to study the hygro-mechanical response of each configuration.

## Materials and Methods

In order to study the hygro-mechanical response of paper due to its interaction with glycerol, three sample configurations are fabricated. Whatman® 3MM chromatography paper with a thickness of 340 um is used to fabricate the samples. A rectangular cantilever beam configuration is selected, with 10 mm of length and 2 mm of width. Three sets of samples are fabricated, one set of wax-free samples and two sets of wax-treated samples. A wax-printer (Xerox® ColorQube 8570) is used to print a wax layer over the samples. An electric oven is used to heat the samples to 115 °C during 8 minutes to melt the wax and distribute it throughout the paper thickness.

A glycerol-distilled water solution at a glycerol concentration of 10 % (v/v) and a pure distilled water solution are prepared in order to study the response of the different samples. An experimental method based on the resonance frequency method is used to determine the hygro-mechanical response of each configuration when interacting with the liquids (Cueva-Perez et al., 2019).

## Results and Discussions

The experimental results are presented in Fig. 1. The resonance frequency of each study case is normalized with the Equation 1 as follows:

$$f_n = f_r / f_0 \tag{1}$$

Where $f_n$ is the normalized resonance frequency at an instant $t$, $f_r$ is the measured resonance frequency and $f_0$ is the resonance frequency of the sample measured before interacting with the liquid. From this figure, it can be observed that pure-distilled water (Water C1) does not alter the dry resonance frequency of the sample in a significant manner. However, a significant difference can be noticed when using glycerol with the same sample type (Glycerol C1), reaching a final value for $f_n$ = 0.87. This result could indicate that a certain portion of solution mass does not leave the paper sample producing an apparently permanent decay in the resonance frequency. The study cases Glycerol C2 and C3 behave in a similar manner; obtaining final values of the normalized resonance frequency of 0.82 and 0.94 respectively. These observations suggest that the glycerol has a significant influence in the mass domain compared to the stiffness one.

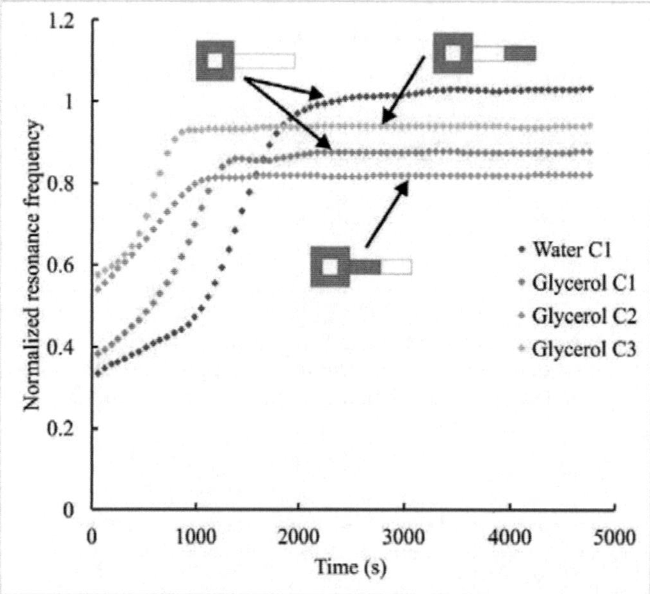


**Figure 1**. Normalized resonance frequency measurements for each study case

## References

*Cueva-Perez I, Osornio-Rios RA, Stiharu I, et al. (2019) Extraction of nonlinear elastic parameters of paper from the amplitude-dependent frequency response of cantilever beams. International Journal of Non-Linear Mechanics 111. Elsevier Ltd: 42–48. DOI: 10.1016/j.ijnonlinmec.2019.01.017.*

*Wang YH, Song P, Li X, et al. (2018) A paper-based piezoelectric accelerometer. Micromachines 9(1). DOI: 10.3390/mi9010019.*

# The Effect of Adding CNT by Direct Compounding Versus Dilution of a Masterbatch on the Elastic and Electrical Properties of PP Copolymer Composites

**Erika Palacios-Aguilar[1*], Jaime Bonilla-Ríos[1], José Antonio Sánchez-Fernández[2], Adriana Vargas-Martínez[1], Jorge de J Lozoya-Santos[1], and Ricardo Ramírez-Mendoza[1]**

[1]*Escuela de Ingenieria y Ciencias, Tecnologico de Monterrey, Ave. Eugenio Garza Sada 2501, Monterrey, N.L., México, 64849.*

[2]*Centro de Investigación en Química Aplicada, Blvd. Enrique Reyna 140, Saltillo, Coah., México, 25294.*

[*]*erikapalacios@tec.mx*

**Abstract**

Composites of multiwall carbon nanotubes (CNT) at 1, 2 and 3 wt.% on a polypropylene-polyethylene random copolymer (PPRC) matrix were prepared by melt compounding CNT powder and by dilution of a commercial polypropylene masterbatch. While the shear viscosity shows similar behavior for both dilution modes, the differences in their elastic properties clearly show the effect of the addition method. This also indicates the relevance of having a difficult to mix masterbatch to enhance the elongational viscosity of the composites for free wall applications such as fiber spinning and blown film. On the other hand, the 2 and 3 wt.% CNT composites from both addition modes have similar conductive behavior, with values near the semiconductors' range. TEM and SEM images show different states of dispersion for each source of CNT; which are the simplest explanation for the differences found in the rheological characteristics of the prepared materials.

## Introduction

Increasing interest in composites comprising carbon nanotubes (CNT) and polymers has been shown in both the scientific and industrial world. Industrial scale up from several research developments has been delayed because many variables are involved in the correct production of the nanocomposites. Several requirements must be accomplished to obtain the desired characteristics to meet a specific performance of the material, e.g., dispersion, alignment, CNT network within the matrix, etcetera. The use of a commercial masterbatch can be an option for the industry to produce polymer CNT composites in high quantities and ideally with secure processing and good dispersion. Therefore, this work is focused on the comparison of nanocomposites obtained from direct compounding of a commercial CNT and a commercial masterbatch.

## Methods/Model Formulations

A PP random copolymer (PPRC) for medical applications (TOTAL Petrochemicals, USA) with a melt flow index of 2 g (10 min.)$^{-1}$ at 230°C and a density of 0.900 kg m$^{-3}$ was selected as the matrix. Two sources of filler were used: a PP masterbatch (Plasticyl PP2001 with NC7000 by Nanocyl™) with a CNT loading of 20 wt.% and CNT (NC7000 by Nanocyl™) with an average nanotube diameter of 9.5 nm, a mean length of 1500 nm, and a surface area of 250–300 m$^2$ g$^{-1}$. Three weight loadings were chosen for this study: 1, 2, and 3 wt.% CNT. Samples from both sources were prepared by a 3-head mixer (ATR Plasti-Corder Brabender GmbH & Co.) at 200°C and 50 rpm for 30 min.

## Results and Discussions

The dynamic rheology of the samples shows a transition from liquid-like to solid-like behavior at approximately 2 wt.% loading of CNT obtained by direct compounding can be observed; data of composites from the masterbatch dilution do not show this transition.

Shear viscosity presents a non-Newtonian behavior in the whole shear rate range. Shear thinning appears at high shear rates, which indicates the formation of networks (polymer-polymer and polymer-nanotube), and an increase in polymer-

CNT interactions. The values of shear viscosity do not change significantly in composites from direct CNT compounding, suggesting a better dispersion when the filler is added directly. The elongational viscosity results (Fig. 1) indicate an enhancement in entanglements promoted by the CNTs in the direct compounded samples and enhanced elastic properties for the masterbatch composites.

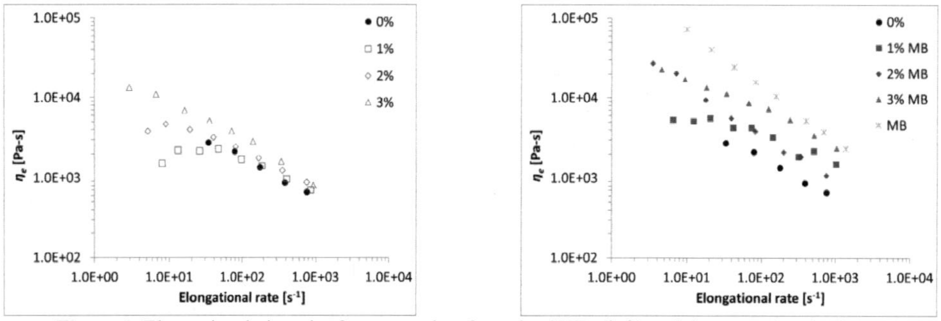


**Figure 1**. Elongational viscosity for composites from the CNTs (left) and the masterbatch (right).

Table 1 shows the resistivity and conductivity values for all prepared samples and some other common materials. 2 and 3 wt.% samples from both nanotube sources are conductive in the semiconductors' range. The difference in 3 wt.% samples leads to the conclusion that the composites from the masterbatch dilution present a higher resistivity because the CNTs are well wetted by the masterbatch PP, making it difficult to form a CNT network within the matrix and maybe even hindering the tunneling effect.

**Table 1**. Electrical test results for all samples under study.

| Sample | Resistivity (ρ) Ω cm | Conductivity (σ) S m$^{-1}$ |
|---|---|---|
| 0% | Overflow | Negligible |
| 1% | $3.17x10^6$ | $3.16x10^{-5}$ |
| 2% | $2.34x10^5$ | $4.27x10^{-4}$ |
| 3% | $9.91x10^4$ | $1.01x10^{-3}$ |
| 1% MB | Overflow | Negligible |
| 2% MB | $1.96x10^5$ | $5.10x10^{-4}$ |
| 3% MB | $2.94x10^5$ | $3.40x10^{-4}$ |
| MB | $6.63x10^1$ | 1.51x100 |
| Silicon | - | $1.60x10^{-5}$ |
| CNT | - | $10^5$ - $10^8$ |
| Copper | - | ~$59.6x10^6$ @ 20°C |

SEM and TEM images show a more dispersed filler in the direct compounded samples compared with those samples from the masterbatch dilution. Agglomerates appreciated in the images of the masterbatch samples look like the masterbatch PP encapsulates them instead of promoting the entanglement with one another.

**References**


*Al-Saleh MH. (2015) Effect of processing conditions on the dispersion, electrical, and mechanical properties of carbon nanotube/polypropylene nanocomposites. Journal of Reinforced Plastics and Composites. 34(9): 742-749.*

*Bikiaris D. (2010) Microstructure and properties of polypropylene/carbon nanotube nanocomposites. Materials. 3: 2884-2946.*

*Boronat T, Garcia-Sanoguera D, Pascual J, Peris F and Sanchez-Nacher L. (2012) Comparative study of the rheological behavior of multiwalled carbon nanotubes and nanofiber composites prepared by the dilution of a masterbatch of polypropylene. Journal of Applied Polymer Science. 126: 1044-1052.*

*Kumar SK and Krishnamoorti R. (2010) Nanocomposites: Structure, phase behavior, and properties. Annual Review of Chemical and Biomolecular Engineering. 1: 37-58.*

30th International Conference on Adaptive Structures and Technologies

# Polymer Gels: Investigation of the Swelling Induced Deformations with a Thermodynamically Consistent Chemo-Mechanical Model

**Marco Rossi[1*], Paola Nardinocchi[2], and Thomas Wallmersperger[1]**

[1]*Institute of Solid Mechanics, Technische Universität Dresden, 01062 Dresden, Germany*

[2]*Dept. of Structural and Geotechnical Engineering, Sapienza Università di Roma, 00184 Roma, Italy*

[*]*marco.rossi@tu-dresden.de*

## Introduction

An increasing demand for adaptive structures using smart materials has emerged in the last decades. These innovative materials are able to meet the required performance of the up-to-date devices improving current systems, especially in the fields of medicine and aerospace.

Among adaptive structures, polymer gels are gaining a particular interest due to their similar behavior to that of biological systems. Polymer gels are elastic materials soaked in a fluid that undergo swelling or shrinking when triggered by an external stimulus such as electric field, pH, humidity, temperature. During the swelling or shrinking of the gel, the fluid is absorbed or released, respectively, while large deformations (40-80%) occur.

The main applications involving polymer gels are bioinspired microstructures, biological tissues, but also fuel cells and actuators. The effectiveness of polymer gels for adaptive structure applications critically depends on the capability of the gel to achieve both prescribed changes in shape and size within the range of requested performance. Currently, several approaches to perform the shape control of swellable materials are pursued for materials in the form of thin sheets ($\sim 10^{-4}$m). Furthermore, frequent swelling-deswelling cycles of the gel may generate critical stresses, fracture, and eventually fatigue issues. It is worth noting that fatigue is still a big challenge for polymers.

## Model Formulation

In the present work, a fully-coupled chemo-mechanical model is presented to investigate the swelling or the shrinking of polymer gels when stimulated by a change in the humidity conditions. The polymer solvent mixture is viewed as single homogenized continuum body. Two fields are taken into account (i) the diffusion of the solvent into the polymer network, and (ii) the swelling induced large deformations modeled through the non-linear mechanics theory. The model is three-dimensional, time dependent, and formulated in a consistent way with the second law of thermodynamics. The derived chemo-mechanical model is investigated analytically by considering a particular setup that allows some approximations, see Fig. 1. The gel, which lays in equilibrium in dry conditions $B_d$, is stimulated (i) by the deformation process $\mathbf{F}_o$ in free conditions, and then is stimulated (ii) by the deformation process $\mathbf{F}$ when surrounded by linear springs. In the latter case, the springs mimic the effects of external bodies like in a single membrane of a fuel cell stack.

## Results and Discussions

The presented model is capable of well predicting the chemo-mechanical behavior of gels (e.g. stresses, stretches, solvent concentration and fluxes) when hydration-dehydration cycles take place due to the deformation processes $\mathbf{F}_o$ and $\mathbf{F}$. The stretch and the stress state of the gel are analyzed with respect to the stiffness of the springs and the external humidity conditions.

Finally, once the stressed configuration B is identified, an incremental theory for the stress-diffusion problem is developed according to the process $\mathbf{F}_s$ by assuming small stimuli and, hence, small deformations: $B \approx B_s$.

This approach is a first attempt to extend the classical Biot's linear theory of poroelasticity for all that kinds of problems where a prestressed porous elastic body swells or shrinks due to an external stimulus.

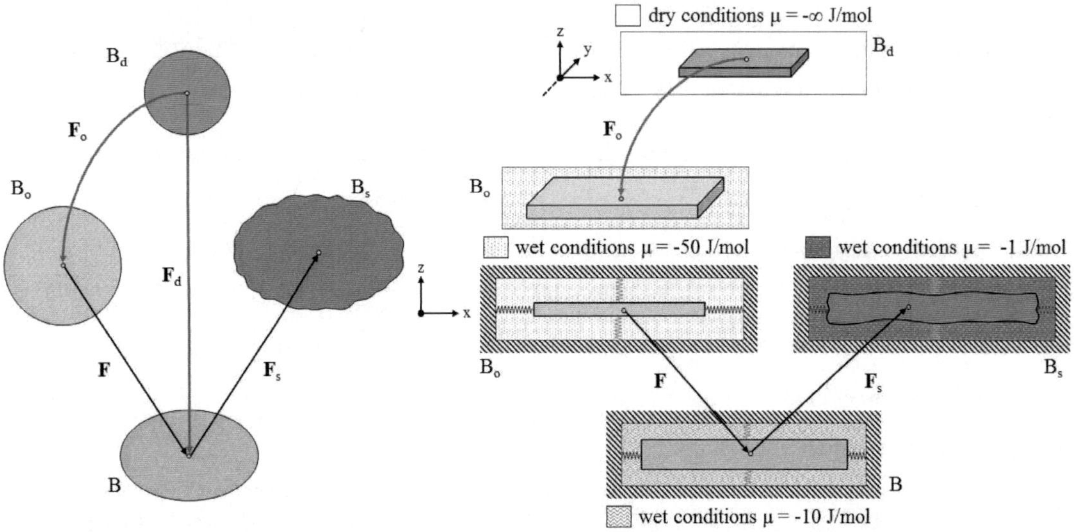


**Figure 1.** The deformation process $\mathbf{F}_o$ brings a dry polymeric gel $B_d$ toward a freely swollen state $B_o$. Then, the membrane is constrained through a uniform distribution of linear springs. A further change in the chemical potential μ induces a hydration-dehydration process and therefore swelling-deswelling of the membrane. The resulting deformation process **F**, brings the membrane towards a new stressed state B. Finally, a small perturbation $\mathbf{F}_s$ of the pre-stressed state identify the configuration $B_s$.

## Outlook

In the future investigation, the authors aim at exploring different working conditions of polymer gels by performing complete numerical simulations of the presented chemo-mechanical model. Moreover, other fields (e.g. electrostatics, ion transport, magnetism) may be implemented in order to describe new effects. It is worth noting that the extension of the proposed thermodynamic approach with different fields can ideally describe the behavior of different adaptive structures.

## References

*Bai R, Yang J and Suo Z (2019) Fatigue of hydrogels. European Journal of Mechanics-A/Solids 74: 337-370.*

*Biot MA (1941) General theory of three-dimensional consolidation. Journal of Applied Physics 12(2): 155-164.*

*Doi M (2009) Gel Dynamics. Journal of the Physical Society of Japan 78(5): 052001-1-052001-19.*

*Inman DJ (2007) Smart Materials and Structures. In Mechanical Vibration: Where do we Stand?, Springer: 139-149.*

*Lucantonio A and Nardinocchi P (2012) Reduced models of swelling-induced bending of gel bars. International Journal of Solids and Structures 49(11-12): 1399-1405.*

*Rossi M, Nardinocchi P and Wallmersperger T (2019) Swelling and shrinking in prestressed polymer gels: an incremental stress-diffusion analysis. Proceedings of the Royal Society A, submitted.*

*Yoon J, Cai S, Suo Z and Hayward RC (2010) Poroelastic swelling kinetics of thin hydrogel layers: comparison of theory and experiment. Soft Matter 6(23): 6004-6012.*

# Rapid Direct Deposition of Filled Polymer on Non-Planar Surface

**Jean-François Chauvette[1], Nicola Piccirelli[2], and Daniel Therriault[1*]**

[1]*Laboratory for Multiscale Mechanics (LM2), Polytechnique Montréal, C.P. 6079, succ. Centre-Ville, Montréal, QC, Canada*

[2]*Safran Composites, Itteville, France*

**daniel.therriault@polymtl.ca*

**Abstract**

We develop a new conformal additive manufacturing (AM) process for a fast direct deposition inside an Aircraft Engine Fan case. The printing platform is composed of a multinozzle mounted on a 6-axis robot that allows printing on the non-planar surface of the fan case. The printing head enables us to significantly reduce the printing time by extruding up to 26 filaments of materials at a time. The feasibility of the adaption of this approach for the Fused Deposition Modeling (FDM) of thermoplastics is also explored.

**Introduction**

Due to unresolved problems such as aircraft noise pollution (Canada, 2004), the aerospace industry is currently a rich source of potential innovation. Therefore, in this project, we develop an industrial AM approach for rapid deposition of filled polymer material that will be used to produce Multi-functional parts for future aircraft engines.

**Methods**

A developed approach for conformal AM using the multinozzle printing head starts with programming the Fanuc's movement trajectory using the RoboDK software (see Fig.1 for details). The trajectory consists of point-to-point interpolations using Cartesian movement. Then, we use the Fanuc's controller outputs to send the proper signals to the pressure controller in order to extrude fugitive ink (Therriault et al., 2005) in synchronization with the robot's movements.

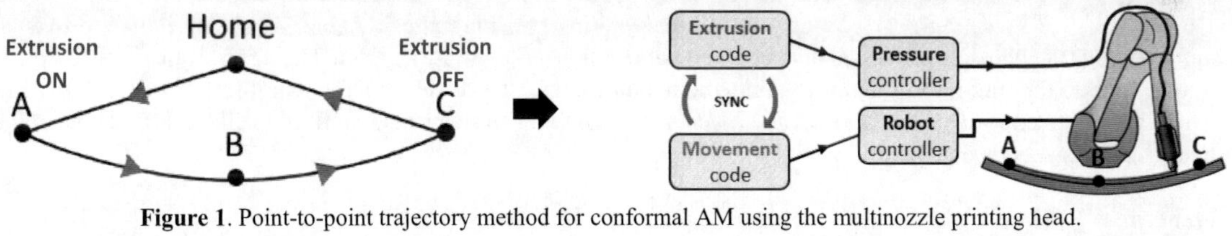


**Figure 1**. Point-to-point trajectory method for conformal AM using the multinozzle printing head.

Additionally, the feasibility for the FDM of thermoplastics is explored using a popular slicing software (Simplify3D). The classical flat sliced toolpath is curved to a desired radius using a methodology described in (Zhao et al., 2018). Figure 2 presents the steps to achieve 3D printing of thermoplastic on a curved surface using a 6-axis robot.

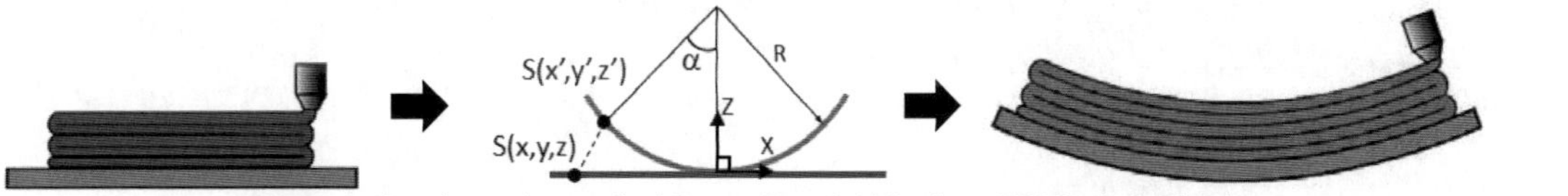


**Figure 2**. G-code curving method for conformal AM using a FDM printing head.

## Results and Discussions

The numerical method for controlling the Fanuc's trajectory toolpath was developed and tested experimentally, as seen in Fig.3. It was observed that, using the proposed simulation tool, the material deposition of one layer of material is feasible at the speed of 15 mm/s.

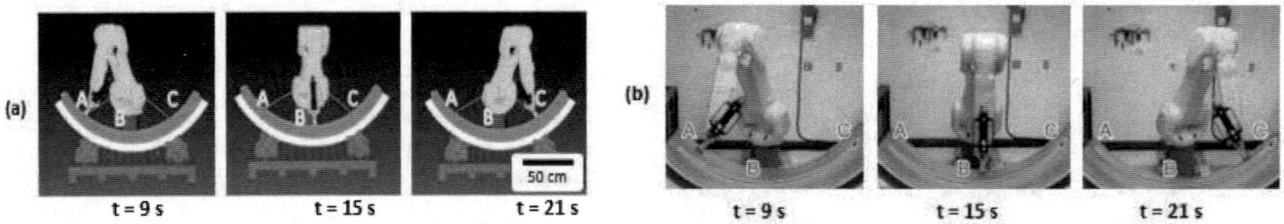


**Figure 3.** Simulation (a) and experimental deposition (b) on the Fan case curved surface using the multinozzle equipped Fanuc robot.

The FDM curved slicing method was implemented and tested with a 3-axis conventional 3D printer (Raise3D pro2). Figure 4 shows the curved slicing steps using a custom-written Matlab script as well as the resulted curved part printed over supports. Due to geometry constraints, a minimum radius of 600 mm can be achieved using the aforementioned printer. With a 6-axis robot, it will be easier to manufacture parts with more complex curvatures. For parts with dimensions of about 90 mm x 36 mm x 30 mm, the curving step takes 30 s. For a Fanuc-compatible part, with dimensions of about 1000 mm x 36 mm x 30 mm, the curving step takes around 120 s, which is suitable for manufacturing purposes.

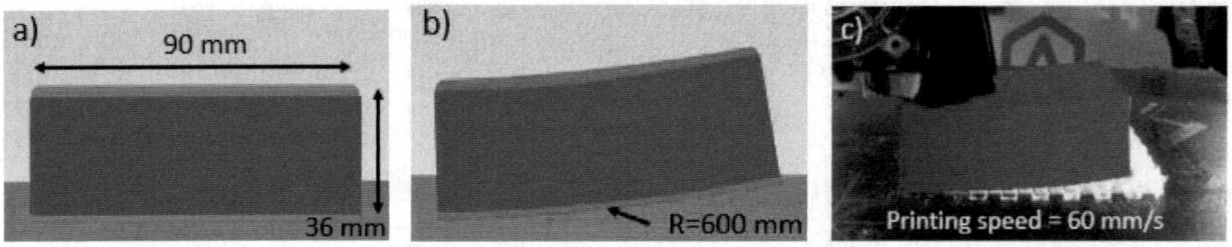


**Figure 4**. Slicing (a), curving (b) and printing (c) of a curved geometry using a 3-axis Raise3D Pro2 printer.

In conclusion, the 6-axis robot reacts as intended to the easily computed toolpath using the RoboDK software. The printability of one curved layer of the fugitive ink inside the engine fan case is proven at 15 mm/s. The FDM approach was tested and a curved part of radius 600 mm was printed using the Raise3D Pro2 printer. For future works, the multinozzle point-to-point methodology is to be developed for multiple layer prints, the FDM approach will be tested with the Fanuc robot and investigation of scanned surface curve fitting will be carried out to obtain the Fan case cross-section equation.

## References

*Canada S (2004) Le bruit des avions près des aéroports. Available at: https://www.canada.ca/fr/sante-canada/services/vie-saine/votre-sante-vous/environnement/bruit-avions-pres-aeroports.html (accessed 29 November 2018).*

*Therriault D, Shepherd RF, White SR, et al. (2005) Fugitive Inks for Direct-Write Assembly of Three-Dimensional Microvascular Networks. Advanced Materials 17(4): 395–399. DOI: 10.1002/adma.200400481.*

*Zhao G, Ma G, Feng J, et al. (2018) Nonplanar slicing and path generation methods for robotic additive manufacturing. The International Journal of Advanced Manufacturing Technology 96(9–12): 3149–3159. DOI: 10.1007/s00170-018-1772-9.*

# Design of a DC-DC Converter Applying Earthquake Algorithm for Inductance Selection

**Efrain Mendez-Flores*, Alexandro A. Ortiz-Espinoza, Israel Macias-Hidalgo, Miguel de J. Ramirez-Cadena, Adriana Vargas-Martinez**, Jorge de J. Lozoya-Santos, Ricardo Ramirez-Mendoza, Arturo Molina-Gutierrez, and Juan C. Tudón-Martínez**

*School of Engineering and Sciences, Tecnologico de Monterrey, Mexico*

**a01336094@itesm.mx, **adriana.vargas.mtz@tec.mx*

**Abstract**

Nowadays in modern industrial applications, such as electric vehicles, space stations, energy harvesting systems, etc., where the power supply efficiency is more important than the output noise performance, DC-DC converters are widely used in order to satisfy the requirements. The efficiency and performance behavior depend on correct parameter estimation for circuits design and operation, where traditional design methodologies cannot ensure an optimal solution regarding the output performance. Nevertheless, component selection and precise estimation of parameters can improve the converter's performance, leading to smaller and more efficient designs. This work, presents and optimization methodology based in a metaheuristic algorithm inspired on earthquakes, for parameter estimation and component selection, using the inductance selection in a non-inverting buck-boost DC-DC converter, designed as supercapacitor charger as a case study. Results validate the proposed algorithm and its implementation selecting optimum inductor value, improving the performance in transitory response with a smaller inductance, also leading to better design and implementation capabilities.

## Introduction

Understanding the relevance of converters in the renewable energies and energy harvesting applications, later this work takes as inspiration the charging method for a supercapacitor proposed in (Chan, Chau, and Chan, 2005), where a DC-DC non-inverting buck-boost converter topology was used for the task. Nevertheless, the proposed design method in this paper seeks to improve it by a smarter method for inductance selection; which uses a novel metaheuristic algorithm first presented in (Mendez et al., 2018) based in earthquake's behavior.

## Model Development

Despite the different topologies and models of DC-DC converters, this document will mainly focus on a smart development of a non-inverting buck-boost DC-DC converter, operating in discontinuous conduction mode (DCM), for a supercapacitor charger application. A similar approach of charging method is discussed in (Chan, Chau, and Chan, 2005), where the same circuit topology was used for the charging purpose, however the approach used a simple mathematical model for the design, without contemplating how to modify the design according to the use of another capacitance, how to find a load current suitable for the load, or even how to select the suitable inductor for the task. The above, motivated to propose a design method that seeks the appropriate inductance for the selected topology for a given capacitance, optimizing its response modeled in the state space approximation (which is also developed in this work).

## Case Study

In order to have a smarter converter design, it is defined for this case study that the main purpose of the circuit is to charge with a better behavior a given supercapacitor. Now, in order to achieve the task, there are some restrictions that have to be considered: (1) According to (Chan, Chau, and Chan, 2005) the voltage of the supercapacitor should never exceed its nominal rated voltage, (2) In this approach, the model considers a fixed frequency and duty cycle for the input, (3) The inductor is the component responsible for the energy transference to the supercapacitor, so there, lies the importance of its correct selection, and (4) With a greater inductance in the circuit, the fly-back effect increases, hence the semiconductors suffer greater wear. The improvement of the implementation was analyzed with the integral of time multiplied by absolute error index (ITAE) taken from (Ogata, 2016), which equation is given by:

$$ITAE = \int_{0}^{T} t \cdot |e(t)| dt \quad (1)$$

## Results and Discussions

Model validation was made in order to prove that the proposed model achieved the expected behavior, and the obtained charging profile (shown in Fig. 1) had a better evaluated absolute behavior. Also, the currently used cost function was selected in order to have a better transitory response in the supercapacitor voltage, without neglecting that an overshoot would be dangerous in a real implementation, under the premise that the nominal voltage of the supercapacitor should not be exceeded, same that was correctly achieved. The obtained inductance value was smaller than the original proposed by (Chan, Chau, and Chan, 2005), where obtaining a smaller inductance improves the implementation capacity of the charger, because smaller physical circuits can be made, with lower cost, and with less electromagnetic charge to the selected semiconductors; which clearly could lead to smarter manufacturing designs, under specific needs of charger specifications.

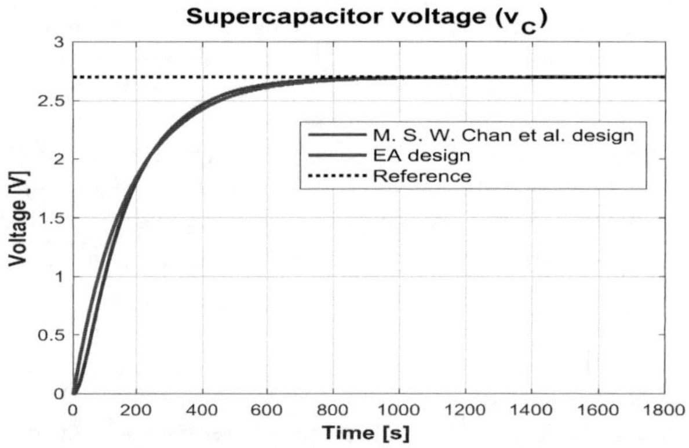


**Figure 1**. Transient response of the Supercapacitor's voltage.

## References

*Arunkumari, T. and Indragandhi, V. (2017). An overview of high voltage conversion ratio DC-DC converter configurations used in DC micro-grid architectures. Renewable and Sustainable Energy Reviews, 77, pp.670-687.*

*Somkun, S., Sirisamphanwong, C. and Sukchai, S. (2015). A DSP-based interleaved boost DC-DC converter for fuel cell applications. International Journal of Hydrogen Energy, 40(19), pp.6391-6404.*

*Chan, M., Chau, K. and Chan, C. (2005). Effective Charging Method for Ultracapacitors. Journal of Asian Electric Vehicles, 3(2), pp.771-776.*

*Mendez, E., Ortiz, A., Ponce, P. and Molina, A. (2018). Electric machines control optimization by a novel geo-inspired earthquake metaheuristic algorithm. In: 2018 Nanotechnology for Instrumentation and Measurement (NANOfIM). Mexico: IEEE, pp.1--6.*

*Ogata, K. (2016). Modern control engineering. [Delhi]: Pearson.*

**Gap in pagination due to unavailable papers.**

**Pages 159-160**

# Compressive-mode Piezoelectric Energy Harvesting

**Zhengbao Yang***

*Department of Mechanical Engineering, City University of Hong Kong, 83 Tat Chee Ave, Kowloon Tong, Hong Kong, China*

**zb.yang@cityu.edu.hk*

**Abstract**

This extended abstract mainly summarizes the research on energy harvesting in the Smart Transducer and Vibration Laboratory at the City University of Hong Kong.

**Introduction**

With the rapid advances in wireless sensors, implantable electronics and wearable devices, the demand for high-power-density and long-lifespan power sources is becoming increasingly stronger. Energy harvesting, emerging as an alternative energy solution to batteries, holds great potential to achieve self-powered autonomous operations of such low-power electronic devices, and thus has attracted much attention from both academia and industry recently. The piezoelectric effect is widely adopted to convert mechanical energy to electrical energy, duo to its high energy conversion efficiency, ease of implementation and miniaturization. The STVL lab at CityU [https://www.cityu.edu.hk/mne/stvl/] has conducted extensive investigations on vibration energy harvesting using piezoelectric materials.

**Methods/Model Formulations**

Piezoelectric energy harvesters, PEHs for short, widely employ the cantilever beam structure with piezoelectric elements attached on the beam surface. The cantilever PEHs suffer from three major problems: low power output, narrow bandwidth and uni-directional sensitivity. Their power output is usually under 1 mW, which cannot meets the power requirement of most potential applications. More seriously, cantilever PEHs work efficiently only around the systems' resonant points. Small variations in the excitation frequency would render the system off the resonance state and lead to a significant drop of power output. Such a poor environmental adaptability obviously cannot meet the requirements from the frequency-variant, multi-directional and random vibrations in the surroundings.

**Results and Discussions**

To address the problems existing in PEHs, we have closely analyzed the piezoelectric material properties and proposed the compressive mode PEHs (Z. Yang, 2014, 2015, 2016; T. Li, 2018). The compressive mode is much stronger than the bending mode and the tensile mode, and thus enable PEHs to have higher performance. Fig. 1 shows the schematic, the working principle, and its theoretical analysis and experimental responses. The maximum power output is 54 mW under a harmonic excitation of 0.5 g, 25 Hz. The prototypes show wide nonlinear operational bandwidth and a capability of harvesting vibration energy from multiple directions. Furthermore, we studied the SSHI energy harvesting circuit and the PMN-PT single-crystal materials. We discussed the energy conversion efficiency (Z. Yang, 2017) and review recent energy harvesting research (Z. Yang, 2018).

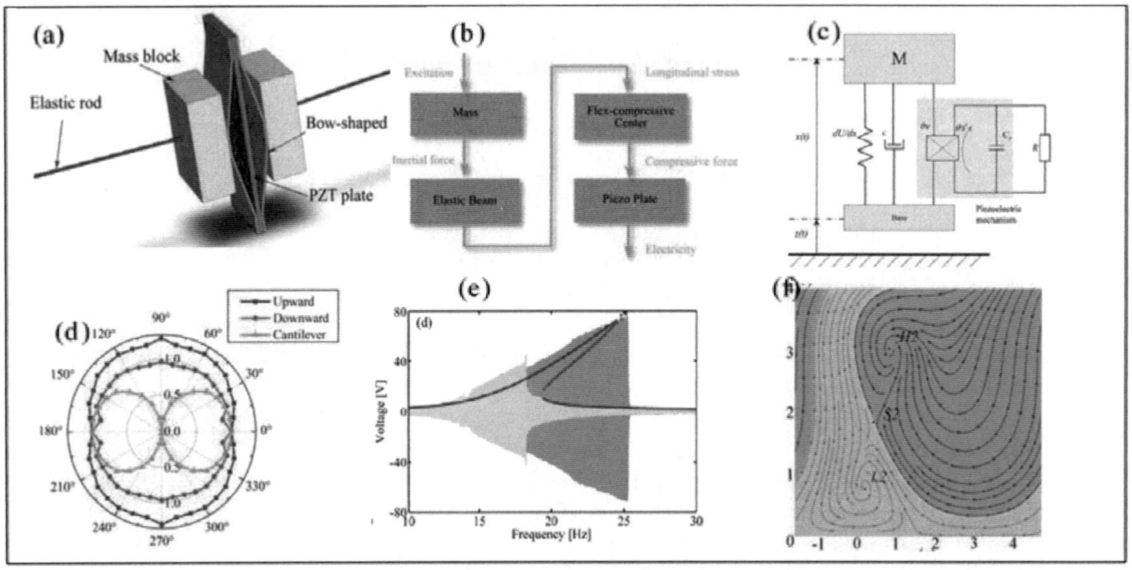


**Figure 1**. Compressive-mode piezoelectric energy harvester. (a) Schematic; (b) working principle of the two-stage force amplification; (c) single degree of freedom Lumped-parameter model; (c) multi-directional sensitivity; (e) voltage responses, 0.5 g, 100 kΩ; (f) reversible hysteresis phenomenon shown by the phase portrait coupled with basin of attraction.

## References


*Haitao Li, Jean Zu, Weiyang Qin and Zhengbao Yang (2018) Modeling and experimental validation of a buckled compressive-mode piezoelectric energy harvester. Nonlinear Dynamics 92(4): 1761-1780.*

*Zhengbao Yang and Jean Zu (2014) High-efficiency compressive-mode energy harvester enhanced by a multistage force amplification mechanism. Energy Conversion and Management 88: 829-833.*

*Zhengbao Yang, Yang Zhu and Jean Zu (2015) Theoretical and experimental investigation of a nonlinear compressive-mode energy harvester with high power output under weak excitations. Smart Materials and Structures 24: 025028.*

*Zhengbao Yang and Jean Zu (2016) Toward harvesting vibration energy from multiple directions by a nonlinear compressive-mode piezoelectric transducer. IEEE/ASME Transactions on Mechatronics 21(3): 1787-1791.*

*Zhengbao Yang, Jean Zu and Zhuo Xu (2016) Reversible Nonlinear Energy Harvester Tuned by Tilting and Enhanced by Nonlinear Circuits. IEEE/ASME Transactions on Mechatronics 21(4): 2174-2184.*

*Zhengbao Yang, Alper Erturk and Jean Zu (2017) On the efficiency of piezoelectric energy harvesters. Extreme Mechanics Letters 15: 26-37.*

*Zhengbao Yang, Shengxi Zhou, Jean Zu, and Daniel Inman (2018) High-Performance Piezoelectric Energy Harvesters and Their Applications. Joule 2(4): 642-697.*

# Importance of Cross-sectional Model in Piezoelectric Beam-type Energy Harvester Analysis

**Shreya Banerjee[1*], P M G B Asdaque[2], and Sitikantha Roy[2]**

[1]*Department of Mechanical Engineering, BMS Institute of Technology and Management, Bangalore-560064, India.*

[2]*Department of Applied Mechanics, Indian Institute of Technology Delhi, New Delhi- 110016, India.*

[*]*shreyamech@bmsit.in*

**Abstract**

Most of the models existing for beam type piezoelectric energy harvester are limited by the fact that they do not provide a generalized definition of beam stiffness and are unable to capture all the classical and non-classical structural couplings (extension-twist, extension-bending, bending-twist etc) arising due to material anisotropy, the effect of arbitrary cross-sectional geometry or coupled electro-mechanics. This may be because of oversimplification due to plane stress assumption, which neglects the effect of off-diagonal terms in the cross-sectional stiffness matrix necessary to capture structural coupling effects. In this present study, we derive a cross-sectional model for both thin and thick beam type energy harvesters, which takes 3D material property as its input without making any simplifying plane stress or plane strain assumption. It gives the flexibility to study material anisotropy, the effect of structural coupling, and arbitrary cross-sectional geometry all in a single common platform. An accurate beam cross-sectional model also help to study the critical design parameters which affect a piezoelectric energy harvester's performance.

**Introduction**

Variational Asymptotic Method (VAM) based analysis provides a generalized modelling platform where any 3D elasticity problem for a beam like structure is divided into a cross sectional minimization problem and 1D beam problem. Yu and Hodges [1] have discussed the merits of VAM which is numerically capable of reproducing the same result using a beam model as one obtains from three dimensional elasticity but with less computational cost. Hodges [2, 3] first applied VAM to derive the cross sectional properties of a passive prismatic beam. Cross sectional analysis of composite beams using VAM was introduced by Cesnik et al. [4]. Roy et al. [5] developed the theory for electromechanical structures which was further extended by Banerjee [6] in the domain of energy harvesters. We have elaborated on that through a comparative theoretical study using six different piezoceramics commonly used in harvester fabrication and subsequently a parametric study is done on material parameters, elastic compliance, piezoelectric coupling coefficient and permittivity.

**Model Formulations**

The coupled constitutive equations which relates 1D resultant forces over the harvester cross section with the 1D strains and voltage are given as [9],

$$\tilde{F} = \frac{\partial H}{\partial \varepsilon}, \quad \tilde{F} = S\varepsilon - \tilde{e}^{a}\phi \tag{1}$$

And,

$$q = \frac{\partial H}{\partial \phi}, \quad q = -\varepsilon^{\mathrm{T}}\tilde{e}^{a} - c_{p}\phi \tag{2}$$

Eqs. (1) and (2) represents the coupled cross-sectional equations for the energy harvester model. To analyze slender energy harvester zeroth-order electromechanical enthalpy is considered and for shear mode or thick beam energy harvester second order asymptotically correct electromechanical enthalpy is considered.

### Results and Discussions

Six different piezoceramics which have been used for this comparative study and model prediction are PZT-5A, PZT-5H, PMN-PT(30%), PMN-PT(33%), PMN-PT and PZT-7A. PZT 7A has the lowest compliance value. For the sake of estimating the performance of all the bimorphs and its effect due to elastic compliance, an average value of compliance and density has been calculated and assumed for all the bimorphs. Because of this the short circuit and open circuit resonance frequency for all the piezoceramics becomes nearby. Performance of two cross-sectional arrangements series and parallel connection has been obtained in Fig. 1. Apart from the elastic compliance and density the piezoelectric coupling coefficient and permittivity have been kept to be in its original value. Power curves have been obtained at the resonance frequency. The maximum power obtained for both parallel and series cross-section is from PMN-PZT bimorph for a wide range of load resistance and the minimum is obtained for PZT 5A.

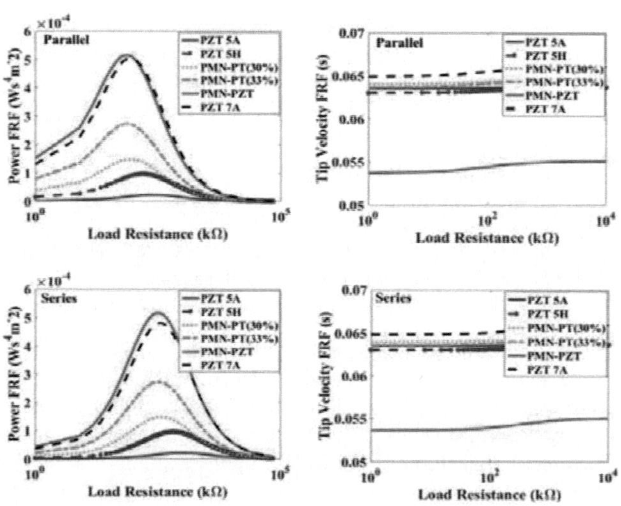


**Figure 1**: Parallel and series configuration power and tip velocity curves for at constant elastic compliance for all six bimorphs made of six different materials

### References

*[1] Yu, W. and Hodges, D.H., 2004. Elasticity solutions versus asymptotic sectional analysis of homogeneous, isotropic, prismatic beams. Journal of Applied Mechanics, 71(1), pp.15-23.*

*[2] Hodges, D.H., 2006. Nonlinear composite beam theory. American Institute of Aeronautics and Astronautics.*

*[3] Hodges, D.H., Atilgan, A.R., Cesnik, C.E. and Fulton, M.V., 1992. On a simplified strain energy function for geometrically nonlinear behaviour of anisotropic beams. Composites Engineering, 2(5-7), pp.513-526.*

*[4] Cesnik, C. and Palacios, R., 2003. Modeling piezocomposite actuators embedded in slender structures. In 44th AIAA/ASME/ASCE/AHS/ASC Structures, Structural Dynamics, and Materials Conference (p. 1803). [5] Roy, S., 2007. A variational asymptotic methodology of smart slender structure modeling. Utah State University.*

*[6] Banerjee, S. and Roy, S., 2018. A dimensionally reduced order piezoelectric energy harvester model. Energy, 148, pp.112-122.*

# Power Output Enhancement of Piezoelectric Energy Harvesters using Electrode Coverage Optimization

**Peyman Hajheidari*, Ion Stiharu, and Rama Bhat**

*Department of Mechanical, Industrial and Aerospace Engineering, University of Concordia, Montreal, H3G 1M8, QC, Canada*

**peyman.hajheidari@concordia.ca*

**Abstract**

This paper reports the theoretical optimization of the active electrode area of a bimorph cantilevered piezoelectric energy harvester connected in series to maximize output power.

**Introduction**

The need of power generation improvements has pushed forward the research to find ways to build piezoelectric energy harvesters (PEHs) with higher efficiency. Shape optimization or geometry modification represents an effective approach as it is possible to enhance the deflection of the free end of cantilever beam (Baker et al., 2005). Although different studies on the non-uniform PEHs have been carried out the effects of electrode coverage optimization on the tapered beams have not been taken into consideration. In this regard, the optimal area of the active piezoelectric layer for a non-uniform beam is evaluated in this paper. Owing to the nature of cantilever beams the volumetric strain is higher near the root of the beam (clamped end) while vibrating (Du et al., 2015). The non-uniformly distributed strain phenomenon can yield an optimal value for the area of the electrodes. Hence, there is a possibility to improve the performance while using smaller amount of electrode material. Following this idea, it was proved that 44% partial electrode coverage starting from the root of the beam where the stress is above 31% of the peak value represents 45% of the total power output.

**Theoretical Modeling**

In this section, the theoretical modelling of the cantilever beam with non-uniform width, $\boldsymbol{b(x)}$ is presented in Fig. 1. The aim is finding the variable $\boldsymbol{x_l}$ resulting in maximized output power.

With the assumptions of Euler-Bernoulli beam theory, the linear piezoelectric constitutive equations (Leo, 2007), under harmonic excitation yield the discretized electromechanical coupling equations for a typical finite element as is summarized in Eq. (1).

$$v(t) = -i\omega\left(i\omega\overline{C_p} + \frac{1}{R_l}\right)^{-1}\boldsymbol{\theta}^T\left[-\boldsymbol{M}\omega^2 + i\omega\boldsymbol{C} + \boldsymbol{K} + i\omega\left(i\omega\overline{C_p} + \frac{1}{R_l}\right)^{-1}\boldsymbol{\theta}\boldsymbol{\theta}^T\right]^{-1}\boldsymbol{F}e^{i\omega t} \quad (1)$$

where $\boldsymbol{M}$, $\boldsymbol{C}^e$, $\boldsymbol{K}$, $\boldsymbol{\theta}$, $\boldsymbol{F}$ are respectively the mass matrix; the damping matrix; the elastic stiffness matrix; the electromechanical coupling matrix and the vector of dynamic forces corresponding to the piezoelectric cantilever beam, and $\overline{C}_p$ is the equivalent capacitance.

**Results and Discussions**

At this stage of study, the effects of electrode coverage ratio on the optimal load resistance and the output power are analyzed, as shown in Fig. 2. The properties of the studied bimorph PEH is summarized in Table 1. In this study, the electrical resistance is considered to be 1000 $\boldsymbol{\Omega}$ and the damping ratio is assumed to be 1 %. It is noticeable that for each active electrode area, the load resistor is varied to find the value matching the internal impedance (Sood, 2003). The results prove that by increasing the ratio of the electrode coverage, the output power increases to hit a peak (close to 50%) and decrease subsequently. For diverging beams the amount of scavenged output power is significantly higher than the uniform and the converging beams. The amount of scavenged power for half-covered diverging beams is 60% higher than the one of the fully covered states. Fig. 2 also depicts that the resistance has an inverse relationship with the electrode coverage ratio.

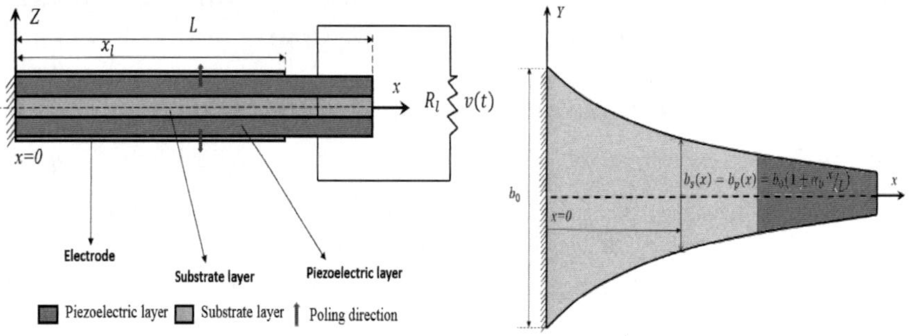


**Figure 1**. Schematic diagram of tapered bimorph piezoelectric energy harvester connected in series configuration.

**Table 1**. Properties of the bimorph piezoelectric cantilever beam

| Property | Piezoelectric layer | Substrate layer |
|---|---|---|
| Material | PZT-4 | Steel |
| Density $\rho$ (kg/mm3) | 7960 | 7800 |
| Length (mm) | 60 | 60 |
| Thickness (mm) | 0.6 | 0.7 |
| Initial width $b_0$(mm) | 10 | 10 |
| Young's modulus E (GPa) | 81.3 | 200 |
| Piezoelectric strain coefficient $d_{31}$(pC/N) | -123 | – |
| Dielectric permittivity at constant strain $\varepsilon_{33}^S$ (nF/m) | 13.059 | – |

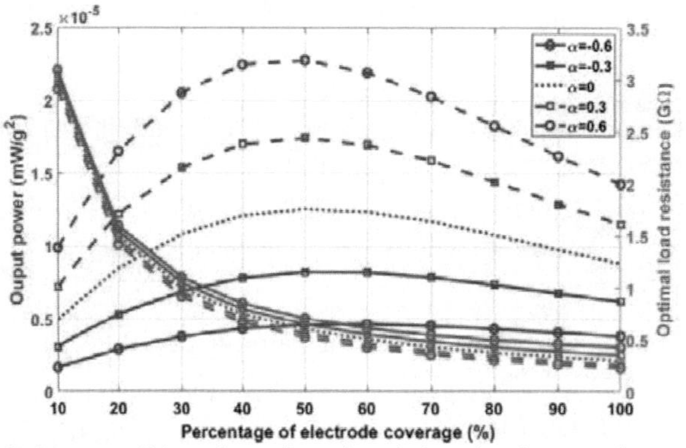


**Figure 2**. An array of bimorph cantilever-based piezoelectric energy harvesters.

## References

*Baker J, Roundy S and Wright P. (2005) Alternative geometries for increasing power density in vibration energy scavenging for wireless sensor networks. 3rd international energy conversion engineering conference. 5617.*

*Du S, Jia Y and Seshia A. (2015) Maximizing output power in a cantilevered piezoelectric vibration energy harvester by electrode design. Journal of Physics: Conference Series. IOP Publishing, 012114.*

*Leo DJ. (2007) Engineering analysis of smart material systems: John Wiley & Sons.*

*Sood RK. (2003) Piezoelectric micro power generator (PMPG): A MEMS-based energy scavenger. Massachusetts Institute of Technology.*

# Investigation into Piezo-magneto-elastic Energy Scavenger with Exponentially Tailored Geometry

**Mahdi Derayatifar*, Ramin Sedaghati, Muthukumaran Packirisamy, and Rama Bhat**

*Department of Mechanical, Industrial and Aerospace Engineering, Concordia University,*

*Montreal, QC, Canada*

*m_deraya@encs.concordia.ca*

**Abstract**

This study, explores the effect of geometrically tailored piezo-magneto-elastic energy harvester on fundamental frequency as well as the output power to mass ratio. The oscillation of the harvester happens via excitation of the base. Case of free vibration has been investigated with analytical method and numerical analysis using COMSOL Multiphysics. In accordance with literature, results express that the ratio of generated power to the mass of the device is increased while operating frequency of the system decreases.

**Introduction**

Vibration based energy harvesting technology has evoked immense research interest due to its high efficiency in generating power for low-power electronics and sensors. Several endeavor has been made to improve the ratio of generated voltage to mass of the device by varying geometrical properties (Kim et al., 2010; Matova et al., 2013; Salmani et al., 2015).

From the studies in (Ben Ayed et al., 2014; Jin et al., 2017), it could be interpret that the rectangular beam generate lower strain than trapezoidal and triangular beams. According to (Pradeesh and Udhayakumar, 2018), the inverted tapered piezoelectric beam generate higher power to volume ratio than rectangular beam. In this research, an inverted exponentially tapered piezo-magneto-elastic energy harvester has been investigated.

**Methods/Model Formulations**

Schematic of a bimorph exponentially width varying piezoelectric beam with magnetic tip mass is shown in Fig.1. The magnetic tip mass is in the interaction with another permanent magnet fixed above. The width is varying according to Eq. 1. Employing the material properties in Table.1, the short circuit natural frequencies are obtained by analytical method based on Rayleigh beam theory along with numerical results from COMSOL Multiphysics software. Figure.2 (a) shows the short circuit natural frequencies for different values of tapering parameter, $C$, with same length of $24.53mm$ and width of $6mm$.

$$b(x) = b_0 \exp(-Cx) \tag{1}$$

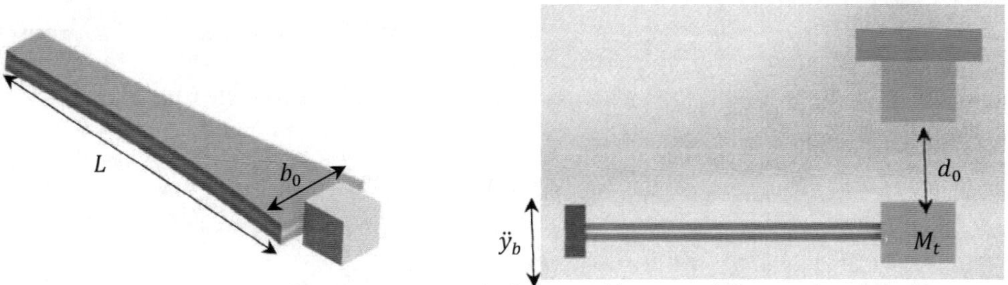


**Figure 1**. Schematic view of the piezo-magneto-elastic beam with exponentially tapered cross section

The magnetic force between the two magnets can be obtained by Eq. 2. Where $\tau$ is the permeability of the free space, $m_1$ and $m_2$ are the moments of the magnetic dipoles, $d_0$ is the initial distance.

$$F_{mag} = k/(d_0 - w(L))^4 \quad , \quad k = -3\tau m_1 m_2/2\pi \tag{2}$$

For the given input acceleration of $0.5g(m/s^2)$ exerted at the fixed end of the beam and the same distance of magnets, $d_0 = 13mm$ and employing load resistance of $R = 100k\Omega$, the generated power to mass ratio with including magnetic force has been obtained for different values of $C$, plotted in Figure. 2 (b). In these figures, it is obvious that $C = 0$ states the rectangular beam.

**Table 1**. Mechanical and electrical properties of bimorph piezoelectric energy harvester

| | Density (kg/m$^3$) | Piezoelectric constant (C/m$^2$) | Permittivity constant (nF/m) | Elastic modulus (GPa) |
|---|---|---|---|---|
| PZT-5H | 7500 | -16.1 | 25.55 | 60.6 |
| Brass | 9000 | - | - | 105 |

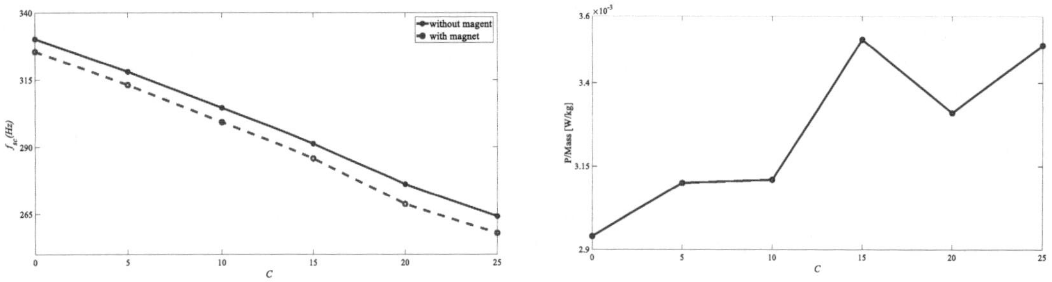


(a) (b)

**Figure 2**. Influence of tapering coefficient, $C$, on (a) short circuit natural frequency, $f_{sc}$ and (b) Generated power to mass ratio, $(P/mass)$.

## Results and Discussions

Due to their higher application in low frequency environments, piezoelectric energy harvesters that are capable of generating higher power while having lower working frequency are more desired. This study shows that compared to the common rectangular piezoelectric beam, inverted exponentially tailored piezoelectric beam have higher power to mass ratio, while possessing lower natural frequency.

## References

*Ben Ayed S, Abdelkefi A, Najar F, et al. (2014) Design and performance of variable-shaped piezoelectric energy harvesters. Journal of Intelligent Material Systems and Structures.*

*Jin L, Gao S, Zhou X, et al. (2017) The effect of different shapes of cantilever beam in piezoelectric energy harvesters on their electrical output. Microsystem Technologies.*

*Kim J, Kim M, Lee H, et al. (2010) Electromechanical characteristics of exponentially tapered Piezoelectric bimorph actuators. Japanese Journal of Applied Physics.*

*Matova SP, Renaud M, Jambunathan M, et al. (2013) Effect of length/width ratio of tapered beams on the performance of piezoelectric energy harvesters. Smart Materials and Structures.*

*Pradeesh EL and Udhayakumar S (2018) Investigation on the geometry of beams for piezoelectric energy harvester. Microsystem Technologies 8. Springer Berlin Heidelberg.*

*Salmani H, Rahimi GH and Hosseini Kordkheili SA (2015) An Exact Analytical Solution to Exponentially Tapered Piezoelectric Energy Harvester. Shock and Vibration.*

# A Frequency Adaptive Package Design for Micro Piezoelectric Energy Harvesters

**Guan-Yu Ke[1], Cheng-Hsiang Hsu[1], Yu-Chun Kuo[1], Wei-Ting Shih[1], Chao-Ting Chen[1], Shun-Chiu Lin[2], and Wen-Jong Wu[1*]**

[1]*Department of Engineering Science and Ocean Engineering, National Taiwan University, Taipei, Taiwan*

[2]*Nano-Electro-Mechanical-Systems Research Center, National Taiwan University, Taipei, Taiwan*

[*]*wjwu@ntu.edu.tw*

**Abstract**

To solve the issue of powering IoTs (Internet of things) devices nowadays, there are more and more researches focusing on scavenging energy from ambient environment. In past years, piezoelectric energy harvester has been widely used for converting kinetic energy to electric energy from any possible ambient sources. Among all energy sources, vibration energy can be obtained in a steady and sustainable form by any machines with motor included. We summarize the energy sources comparing by acceleration levels and vibration frequency of machines shown in Table 1[1][2]. In our past research, we have been successfully demonstrating as an Internet of Things (IoTs) application scenario, a BLE (Bluetooth low energy) temperature sensor continuously transmitting data to mobile application only powered by our self-fabricated piezoelectric energy harvesters (PEHs)[3]. Since the cantilever type micro energy harvester has high quality factor with narrow operating frequency range around resonance, we used to design the natural frequency of PEHs matching to vibration frequency of targeting machine or appliances. However, due to the mounting tolerances of package fabrication, the operation frequency and vibration frequency may not match so well. In this study, we try to propose a package design for our piezoelectric energy harvesters which can adjust the resonant frequency of the cantilever structure adapting to the vibration frequency of vibration sources. To achieve this objective, we design a package with simple adjusting mechanism with screws which helps to adjust the effective length of micro cantilever beam fitting with the mounted vibration sources. The design of the package, the adjustment mechanism, and the testing results will all be detailed in this paper.

**Table 1**. Ambient vibration sources

| Source | $f_1$, $a_1$ | $f_2$, $a_2$ | $f_3$, $a_3$ |
|---|---|---|---|
| Microwave oven top [1] | $f_1$= 120 Hz | $f_2$= 120 Hz | $f_3$= 240 Hz |
| | $a_1$= 1.11 g | $a_2$= 1.11 g | $a_3$= 0.57 g |
| Air conditioner blades [2] | $f_1$= 51.5 Hz | $f_2$= 121 Hz | $f_3$= 312 Hz |
| | $a_1$= 0.18 g | $a_2$= 0.1 g | $a_3$= 0.18 g |
| Motorcycles [2] | $f_1$= 27.1 Hz | $f_2$= 54.4 Hz | $f_3$= 60.3 Hz |
| | $a_1$= 0.19 g | $a_2$= 0.79 g | $a_3$= 0.77 g |